MW00737361

Managing Engineers and Technical Employees

How to Attract, Motivate, and Retain Excellent People

For a complete listing of the *Artech House Professional Development Library*,
turn to the back of this book.

Managing Engineers and Technical Employees

How to Attract, Motivate, and Retain Excellent People

Douglas M. Soat

Artech House
Boston • London

Library of Congress Cataloging-in-Publication Data
Soat, Douglas M.
Managing engineers and technical employees : how to attract, motivate, and retain excellent people / Douglas M. Soat.
p. cm.
Includes bibliographical references and index.
ISBN 0-89006-786-4 (alk. paper)
1. Engineering firms—Personnel management. I. Title.
TA190.S667 1996
96-4673
CIP

British Library Cataloguing in Publication Data
Soat, Douglas M.
Managing engineers and technical employees : how to attract, motivate, and retain excellent people
1. Engineering—Personnel management
I. Title
620'.00683

ISBN 0890067864

Cover design by Darrell Judd

685 Canton Street
Norwood, MA 02062

International Standard Book Number: 0-89006-786-4
Library of Congress Catalog Card Number: 96-4673

10 9 8 7 6 5 4 3 2

This book is dedicated to my wife, Lynn, my sons, Jim and John,
and in memory of my daughter, Jennifer

Note to the Reader

Contents

Acknowledgments

I want to thank Mark Walsh, Kimberly Collignon, Beverly Cutter, Matthew Gambino, Darrell Judd, Laura Esterman, and the other individuals from Artech House who helped to make this book a reality.

I also want to express my sincere appreciation to Lynda Salisbury, who did an outstanding job typing part of the original draft and nearly all of the final manuscript. In addition, I want to thank Cheryl Messina, Lynn Soat, and Jim Soat for their help in typing a portion of the original draft. Jim also had an excellent idea concerning the graphics for the book cover.

I want to express my gratitude to my clients and former employers for giving me the opportunity to develop, refine, and apply various ideas that I present in the book.

I owe an invaluable debt to my mother, Doris Soat, for her support and encouragement while I was growing up. Without this, many of my accomplishments probably never would have been realized, including this book.

Finally, I want to thank my wife, Lynn, and my sons, Jim and John, for their understanding, support, and encouragement while I was engaged in this project.

Introduction

Audience for This Book

This book was written for practicing managers of technical employees. By technical employees, I mean a broad range of skilled professionals, including engineers, computer scientists, chemists, physicists, actuaries, and other individuals who have had specialized education in mathematics or science. These employees may be working in applied fields, such as manufacturing engineering, or they may be employed in more theoretical scientific areas, such as basic research in organic chemistry. Technical employees may also include other types of professionals in the financial, medical, and legal fields.

This book is intended to assist individuals who work in manufacturing and service industries, government, nonprofit organizations, and educational institutions. The target audience is people who have had some experience managing technical employees. However, individuals who have not yet had managerial experience but aspire to do so in the future will also find this book useful and not too difficult to understand.

Managers at several different levels within organizational hierarchies will find this book helpful. That is, it will be useful to first-line supervisors of technical employees, such as project managers or supervising engineers, and be equally valuable to managers at a much more senior level, such as vice presidents of engineering or chief chemists.

A manager who does not have a technical background but the responsibilities involving technical employees will also find this book to be extremely helpful. Such individuals might include people such as presidents, executive vice presidents, general managers, directors of human resources, and executive directors.

University professors and others who teach courses in the management of technical employees will find this book to be a valuable resource for students who want a practical guide to the topic.

A Unique Perspective

My education and experience are unique. Educationally, I not only have a B.A., M.S., and Ph.D. in psychology, but I also have an M.B.A. in management. (Although there may be some others out there, so far I have not met another psychologist who has an M.B.A.) Regarding experience, I have worked as an inhouse professional, a human resources executive, and an external consulting psychologist. I have worked for and with manufacturers, service organizations, nonprofit organizations, educational institutions, and government agencies. This experience has included work with numerous technical and nontechnical employees and as a practicing manager.

This unique background enables me to provide readers with a very different perspective and special insights regarding the topic of the management of technical employees.

Purpose of This Book

Why was this book written? It was completed to serve as a practical guide to individuals who manage technical employees.

Numerous definitions of management can be found in the literature [1]. Some are relatively simple and others are more complex. My favorite definition is a relatively simple one: management is getting tasks accomplished through others.

Every conscientious manager, regardless of their organizational level, industry, and staff , needs to be concerned about five key areas. These include attracting, selecting, developing, motivating, and retaining competent, high-performing employees.

Those who manage technical employees are no different; they need to be concerned about these same five areas. Although there are many similarities between technical and nontechnical employees, there are also significant differences. For example, technical employees often tend to be more intelligent, analytical, careful, creative, curious, introverted, self-sufficient, and independent than nontechnical employees [2, 3]. Therefore, the managers of technical employees are faced with special opportunities, challenges, and problems.

This book deals with these distinctive opportunities, challenges, and problems. It provides managers of technical employees with easy-to-understand, practical ideas and techniques that they can apply in order to do an effective job of attracting, selecting, developing, motivating, and retaining top-notch people.

How This Book Is Different From Others

Although I am certain that virtually every reader of this book will find some information in it that they have not seen before, the various points are, for the most part, not totally novel.

My book is based, however, on a premise quite different from virtually every other management book I have seen. Everyone else seems to be touting some "new" approach that is supposed to be the panacea for all of management's problems. This approach fits in nicely with what most managers seem to be seeking: a "quick fix" to their problems.

However, after having had over 20 years of experience as a psychologist and a human resources executive, I am certain that there is no such thing as a quick fix for management problems. Some of the new ideas certainly have some merit, but no approach can help an organization to improve its effectiveness dramatically without allocating a great deal of time and effort.

In addition, 20 years of observing dozens of organizations in numerous different industries has led me to the conclusion that before managers spend any time searching for new approaches, they need to ensure that they have truly mastered the basic principles involved in the recruiting, selection, development, motivation, and retention of employees.

It has been my observation that very few managers have put in the time and effort to master these basics. For example, I have observed that only a small number of managers spend adequate time with staff members discussing specifically what they will do differently after they have completed a particular course. That managers need to spend a considerable amount of time discussing specifically how staff members will use what they have learned in a course is not a new idea. However, it is important and it is timeless; it was true 20 years ago, it is true today, and it will be true 20 years from now.

My intent in this book is to provide technical managers with various practical, basic ideas that may not necessarily be "brand new" or based on the latest research but are valid and important nonetheless. I have provided a substantial number of references to support the long-term validity of these ideas. I believe that any individual who takes the time and effort needed to implement my various suggestions will be a much more effective manager.

Overview of The Book

The remainder of this book deals with five major topics: attracting, selecting, developing, motivating, and retaining excellent technical employees.

Chapter 1 deals with recruiting top-notch technical people. Readers will discover how providing highly competitive wages and benefits and developing a reputation for having outstanding products/services and people can enable a manager to attract excellent technical employees. The effective use of novel recruiting techniques, such as realistic job previews, is also described. Finally, valuable sources of top-flight technical employees, such as formal employee referral programs, are discussed.

Chapter 2 describes the critical process of setting appropriate selection standards. Setting high, yet realistic, standards is discussed. In addition, the reader is shown how to match technical candidates with a given job, supervisor, and corporate culture.

Chapter 3 gives the reader some practical tips on how to interview technical candidates effectively. The importance of planning prior to an interview is described. Some specific practical suggestions on how to conduct an effective interview and some suggested questions to use and avoid are provided. Finally, some practical tips on the use of psychological principles of observation and interpretation are described.

Chapter 4 deals with the proper use of other selection devices with technical employees. These include tests, work samples, reference checks, and assessment centers.

Chapter 5 covers key selection criteria for technical employees, using a case study. The reader is provided with a list of selection criteria for a specific technical position. In addition, examples of questions that could be used to evaluate candidates for this technical position are given.

Chapter 6 deals with one-on-one development. It explains why and how a manager of technical employees must be and can be involved in the entire process of their employees' development. The use of the novel concept of ability development teams is also discussed in detail. In addition, the effective use of "360 degree feedback"–based tools in the development of technical employees is discussed. Some original 360-degree-feedback instruments are described.

Chapter 7 describes some practical techniques that the reader can use in group inhouse training/development. Included is material on such topics as adult learning, effective training needs analysis, training "do's" and "don'ts," proper evaluation of training effectiveness, and types of inhouse training (for example, orienting new employees to the corporate culture).

Chapter 8 deals with means of developing technical employees in addition to those that described in the Chapters 6 and 7. These include outside seminars and courses, formal reading/discussion programs, team building, career development, and succession planning.

Chapter 9 deals with what I believe are the two most critical motivators that a manager can provide for technical employees, that is, setting high, yet realistic, expectations and demonstrating true concern for their welfare. Examples of individuals who successfully demonstrate and who fail to use these two motivators are given. Specific, practical ideas are provided for the manager of technical employees to help them use these two critical motivators effectively.

Chapter 10 deals with other key motivators for technical employees. These include providing interesting and intellectually stimulating work; giving effective feedback; rewarding top performance; eliminating ineffective performance; using peoples' talents; building a true "team"; listening and dealing with problems; communicating thoroughly, candidly, and promptly; and allowing considerable autonomy. Specific examples of effective and ineffective use of these motivators is discussed. In addition, specific, practical ways that a manager of technical employees can take advantage of these key motivators are described.

Chapter 11 explains how a manager of technical employees can successfully retain their best employees. It discusses how using the various techniques discussed earlier leads to increased retention. It also describes how treating all technical employees with respect allows a manager to minimize turnover and keep the very best performers.

The final chapter of the book shows readers how to integrate the various ideas and techniques discussed previously so that they can "do it all" (that is, successfully attract, select, develop, motivate, and retain the very best technical employees available).

References

[1] Koontz, Harold, and O'Donnell, Cyril, *Essentials of Management*, New York: McGraw-Hill, 1974.

[2] Coss, Frank, *Recruitment Advertising*, American Management Association, New York, 1968.

[3] Fear, Richard, *The Evaluation Interview*, Third Edition, New York: McGraw-Hill, 1984.

Chapter 1

Recruiting

1.1 INTRODUCTION

If a technical manager wants to launch or rebuild an organization with outstanding technical employees, the first step is to attract such individuals. You can attract top-flight people to your organization by making employees feel that your organization is a desirable place to work. The key factors that will make an organization attractive are described in this chapter.

In addition, there are also some recruiting resources and techniques that can be extremely helpful in a technical manager's efforts to locate excellent technical employees. These are also described in this chapter.

1.2 KEY FACTORS THAT ATTRACT AND AID THE RECRUITING OF OUTSTANDING TECHNICAL PEOPLE

Several factors are critical in making an organization desirable to top-flight prospective technical employees and the recruiting of such individuals an easier job. First, a technical manager needs to provide outstanding compensation. Second, the organization needs to provide an excellent benefits package. Finally, an organization needs to develop a reputation for having highly effective, well-treated employees and top-quality products or services that are on the cutting edge in the company's industry. Each of these factors will be addressed in turn.

1.2.1 Outstanding Compensation: Base Salary

Many managers of engineers and other technical employees make two major erroneous assumptions regarding compensation. They assume that (1) money is not important to technical employees, and (2) they can attract excellent people with moderate pay.

First, money is not unimportant to engineers or other technical employees, any more than it is unimportant to any other employees. Technical employees have families to support, mortgages to pay, and car payments to make just as do nontechnical employees. They enjoy having nice material possessions, going on fun-filled vacations, and putting aside money for future needs (for example, children's educations and retirement), as do individuals in nontechnical occupations. So technical employees cannot afford to ignore financial considerations any more than anyone else can. In fact, since many technical employees are quantitatively oriented, it is common for them to study financial information carefully. For example, they might compare different opportunities with regard to the retirement benefits, relocation costs, and cost of living.

Some of the misconceptions regarding the importance of money to engineers and other technical employees may have arisen as a result of the influence, and perhaps misinterpretation, of the "two-factor theory" developed by Herzberg et al. [1]. According to this approach, compensation is one of the "dissatisfiers" or "hygiene factors." If someone is not paid enough, he may be dissatisfied. Paying a person what he feels he is worth will eliminate the dissatisfaction. However, paying a person more than this will not further "satisfy" or motivate the person.

Thus, according to this theory, proposing less than what a person feels is a fair salary to recruit a technical employee might cause an individual to reject a job offer. However, proposing more than this person's minimally acceptable figure is not likely to affect that individual's willingness to accept the offer.

Herzberg and his colleagues' own research [1] verified his theory. Unfortunately, however, efforts of other researchers to replicate his findings have been largely unsuccessful. Research on the two-factor theory has been rather limited in recent years, according to Saal and Knight [2]. These authors reported that various serious problems have been identified regarding Herzberg's approach (for example, inconsistencies in his description of his two-factor theory) [2].

Many managers of technical employees continue to believe that offering an average salary should be quite sufficient to attract a top-flight person. However, the logic does not make sense. Why would an engineer, for example, who is among the top 5 to 10 percent of performers in her field accept a job offer from an organization that is offering a salary that is just "average" (for example, at the going market rate for "competent" performers)? The answer is, she probably would not.

People in various organizations with which I have worked always say that they want the "best" employees. For example, an executive in one company said that he wanted to attract employees in the "top 10 percent" regarding intelligence and work ethic. Yet this person, when discussing salary offers for these potential employees, said that he did not want to pay more than the "going market rate" for "competent" individuals. This simply will not work.

I believe that an organization needs to be realistic. If a company really wants to attract the top 10 percent of technical employees, it will need to offer salaries that are in the top 10 percent, as compared to other companies. If a company is only willing to pay an "average" salary, then realistically it is only likely to attract "average" performers.

Some managers of technical employees have said they cannot "afford" to pay high salaries. My response to these people is simply that they cannot afford *not* to pay well. High salaries should not be viewed as a "waste of money." Rather they need to be viewed as a "cost of doing business" or an "investment" in talent. Companies recognize that they need to invest considerable capital in plants and equipment if they want to stay on the "cutting edge" and be an industry leader. They need to recognize that paying top salaries is simply another form of investment necessary to maintain a competitive edge.

Salaries are complex motivators because they are not simply an amount of money that can be used to buy material goods. They are also an index of someone's value to an organization and one indicator of someone's level of success in her profession [3]. As a result, if a potential technical employee is offered a high salary, she is more likely to believe that the company *truly* considers her to be an extremely valuable potential asset to the organization. The individual may also feel that she has attained a high level of career success as a result of getting the generous offer. Thus, high salaries can truly help an organization attract excellent technical employees, even if these people are not particularly interested in many of the material possessions that could be bought with a generous salary.

1.2.2 Outstanding Compensation: Bonuses, Stock, and Stock Options

Other aspects of compensation that can help attract excellent technical employees include bonuses and stock or stock options. First, regarding bonuses, most organizations that have bonus plans have components based on individual performance and factors based on unit and/or corporate performance. This type of arrangement is ideal, because if a technical employee makes an outstanding contribution, but his unit or the corporation as a whole has a bad year, the person's exceptional performance can still be rewarded with a bonus.

For example, suppose an R&D engineer in a major telecommunications company invents a new type of fiber optic cable that costs 50 percent less than the existing cable. Suppose also that the new cable works equally as well as, or perhaps better than, the currently used cable and is likely to save the company millions of dollars per year. Because of the significant future impact on the company's "bottom line," this accomplishment should be recognized with a sizable bonus, even if the company has just had a mediocre or even a dismal year financially. The definition of "sizable" has to be determined by each organization, but in general it should be at least somewhat reflective of the anticipated financial impact of the accomplishment. In other words, using the previous example, taking the engineer out to dinner and giving him a gift certificate worth $200 hardly seems appropriate, considering the likely company savings of millions of dollars per year.

What then, is an appropriate bonus? One example of which I am aware concerns the head of an engineering group of a manufacturing firm. The company had a new plant start up, and major problems existed, regarding both capacity and quality. The situation soon reached crisis status, in that a major customer was going to be shut down if the numerous problems were not solved quickly.

The engineering manager implemented various solutions to the problems, such as adding new, more sophisticated equipment with greater capacity and "error-proofing" the manufacturing process. He and his group had to work almost literally day and night, but the problems were solved in time and the crisis was avoided.

The company recognized that were it not for the engineering manager's and his staff's exceptional efforts, a major customer would have been shut down. As a result, the customer's business, worth millions of dollars per year, would probably have been lost. The engineering manager was given a $30,000 bonus, even though the organization had a poor year financially, and no bonuses based on the company's financial performance were paid to anyone else.

In addition to bonuses, a company can also grant stock or stock options to reward employees for exceptional performance. For example, it has been reported that the Microsoft Corporation helped create over 2,000 millionaires among its employees as a result of stock options that were offered to them [4]. The Microsoft Corporation has achieved phenomenal growth and success, and I am confident that providing the stock options to these employees has contributed significantly to the company's success.

1.2.3 An Excellent Benefits Package

The employee benefits package is another important factor that affects recruiting [5]. Providing outstanding benefits is essential if a company wants to recruit excellent technical employees. With benefits costs, particularly in the health care area, continuing to escalate at a rapid rate, many employers are looking for ways to slow or to stop the cost increases. Various methods are used, such as self-insuring for medical costs, aggressively comparing the pricing of insurance coverage, promoting and emphasizing wellness among employees, and precertification of hospitalization.

Although employees may need to share some of the burden of increased costs (for example, by assuming their proportionate share of health insurance increases), employers need to be very careful when considering reduction of benefits. Outstanding prospective technical employees typically analyze the compensation package carefully, and they usually examine a company's benefits package. If a technical manager believes that she can attract the top 10 percent of technical employees, despite having a benefits package that is less than or barely meets the "typical" package, she is deluding herself. Why should an outstanding prospect work for an organization that offers mediocre benefits when companies with excellent packages are also interested in him?

One company with which I am familiar, for example, reduced the company's health benefits several years in succession to save money, even though one year the health insurance premiums did not increase at all, due to favorable claims experience. The president of the company felt that lowering insurance costs was essential, and he erroneously believed that reducing benefits would not have any significant impact on recruiting or retaining current employees.

Unfortunately, the impact on potential or current employees, though unquestionably significant, is often not obvious. For example, when an outstanding candidate rejects a job offer, he is seldom totally candid about the reason(s). In addition, candidates themselves sometimes may not be totally aware of all of the factors that influence a decision.

When an organization has a benefits package that is only mediocre or even subpar, several messages are sent to prospective employees. This is true regardless of whether the messages are intended or not. It is also true even if management statements specifically contradict these messages. For example, an organization may tell prospects that it "really cares about employees," and its literature may claim, "Our employees are our greatest asset." However, when that same company has a mediocre benefits package, the message conveyed to prospects is, "We don't want to spend too much on our technical employees" or "We don't really care all that much about our technical employees and their families' needs." These latter unintentional messages influence prospects more than the company's statements do because prospective employees recognize the validity of old adages such as, "Talk is cheap," or "Put your money where your mouth is."

Conversely, a company that provides a benefits package that is considerably above average for the industry and for the geographical area is reinforcing any statements it makes concerning the importance of technical employees.

Managers in some organizations may say they cannot "afford" above-average benefits packages. Unfortunately, if they truly want to attract top-notch employees, they cannot afford *not* to provide outstanding benefits.

1.2.4 An Outstanding Reputation

In addition to the importance of an excellent compensation and benefits package, one other essential ingredient is needed to attract outstanding employees, that is, an excellent company reputation. The reputation of an organization is critical in two areas: (1) employees and (2) products or services.

First, regarding employees, an organization must have a reputation for having highly talented and well-treated employees. If a company is well known for having extremely talented technical employees, other outstanding technical employees will want to work for that organization. Second, it is critical for an organization to have an outstanding reputation for how it treats its employees. If an organization is well known for its outstanding compensation and benefits, as well as for being a great place to work, this will make recruiting excellent technical employees a much easier task. Conversely, if an organization has a reputation for paying at the market rate, or perhaps somewhat below the going rate, this will make it difficult to attract top-flight people. Similarly, if an organization has a reputation of a "sweatshop," where people put in long hours but get very little in the way of rewards or recognition, this will make it very difficult to attract outstanding people.

Another key aspect of an organization's reputation concerns how it is perceived by people in terms of its products or services. Having a reputation as a true "cutting edge" company will make the process of attracting individuals with outstanding skills much easier. This is particularly true regarding R&D engineers and other technical people who work in the R&D area. For such individuals, being on the forefront of technology is of great interest. Working for a company known for technological innovations in products and services is a great attraction for people who work in R&D because this suggests to them that they may have an opportunity to use all of their creative talents. On the other hand, an organization that has a reputation for not being particularly innovative and following a "me too" strategy in terms of the development of products or services is likely to find it much more difficult to attract R&D people with significant creative abilities.

How does a company build an outstanding reputation with regard to its employees and its products or service? Word of mouth is probably the most significant factor in developing such a reputation. An organization needs to recognize that anyone with whom its employees come in contact represents a potential marketing tool for attracting new employees. If the people who have contact with an organization feel that the company is not a particularly good one, then their influence on attracting top-flight employees will primarily be a negative one. For example, two people may be having a discussion and one asks the other, "Do you know anything about the XYZ Company?" The first person, who has had some type of contact with the XYZ Company, indicates, "Oh yes, I've heard a few things. I have heard that it isn't a very good place to work, because they don't treat their employees all that well." Obviously if the person who heard this message were in a position now or at some future time to be considered for a job at the XYZ Company, the comment that the person heard concerning XYZ's reputation is likely to have some influence on her. One negative comment may not necessarily deter someone from joining an organization. This is particularly true if that person has heard a number of other favorable things about the company, or if he has directly experienced some positive aspects of the organization. However, what every organization must remember is that even one negative comment has the potential to influence a prospective employee in some way. It is possible that some potential employees may not even take the initial step to find out more about an organization if the only comment that they have heard about the company is a negative one. The more credible the source of the statement is, the more likely it is that the comment will have an adverse impact.

Conversely, if a potential employee has heard a positive comment about a particular company, she might take some action, such as trying to find out more about the organization. If this additional information is favorable, the individual might well take some overt action to try to become a part of the organization (for example, she may send in her resume for possible consideration).

As I indicated earlier, anyone who comes in contact with an organization can have a potential impact on that entity's ability to attract top-flight technical employees. Consider, for example, an organization's customers. They obviously are of critical importance. These are the individuals who can determine the success or failure of an organization, depending on whether they are satisfied or dissatisfied with the company's products or services. However, in addition, customers can play a significant role in the recruiting process. Customers have many occasions on which they may make a comment about an organization to others. For example, an individual may make a statement to a friend such as "Yes, I buy all of my cassettes from the ZYX Company. I have to admit I have never gotten a defective product during the ten years that I have been a customer of this organization." Anyone hearing such a comment would undoubtedly be impressed with this company. In addition, this favorable impression would have an influence, perhaps even a significant one, regarding this individual's possible current or future interest in becoming an employee of the organization.

Another critical group is an organization's suppliers. They can be just as significant in terms of affecting a company's ability to recruit top-notch technical employees as the organization's customers can.

Rejected employees represent another group that can affect a company's ability to recruit top-flight technical employees. Suppose, for example, that a person applies for a position and never gets any response whatsoever from the company. This person is likely to have a dim view of this organization and may express this opinion to others who are potential employees. On the other hand, if an individual applies for a position and gets a response in a very short period of time, this is likely to have a favorable affect on his opinion of the company. For example, I have heard individuals make comments such as "Yes, I interviewed with the Acme Corporation. Unfortunately, I didn't get a job there but was really impressed by the way I was treated. Even though the company didn't end up making me an offer, they kept me posted of my candidacy during the process. Also, they were true professionals in all of their relationships with me." If a person hears a comment such as the one just made, she would be likely to think favorably of the Acme Corporation. If this individual ever has an opportunity to be considered for a job at Acme, she is likely to relish the idea of being a candidate for a position there.

Finally, one additional group that is probably one of the most critical in terms of recruiting other top-notch technical employees is an organization's current employees. If the current employees think that a company is a great place to work, they will undoubtedly pass on this opinion to the various people with whom they have contact. Conversely, if current employees are not being treated with respect, it is also likely that they will share their perception of the company with others.

1.3 RESOURCES AND TECHNIQUES THAT AID EFFECTIVE RECRUITING

As I indicated in Section 1.2, providing a superior compensation and benefits package and developing an outstanding reputation are extremely helpful in recruiting excellent employees. There are also numerous resources and techniques that can aid the recruiting process. These will be described in the sections that follow.

1.3.1 Employee Referrals

Current employees can be one of the most valuable resources for helping to find excellent new technical employees. Some research has indicated that some of the best people hired are referred by employees [6]. Employee referral programs can be informal or formal [6, 7]. Informal employee referral programs might simply involve asking employees on a very informal basis to recommend good people that they know. Another approach to informal employee referral might be to post a notice on the bulletin boards indicating that management would greatly appreciate employee referrals of any strong candidates for available openings.

An organization may also implement a formal employee referral program. Such a program might typically involve a specific procedure that describes how employees should refer appropriate candidates to management for openings. One key aspect of a formal employee referral program is providing some sort of incentive. Many employees may be happy to refer good people for openings simply as the result of knowing that management would appreciate such referrals. However, if some type of specific reward is provided to employees who refer excellent prospects for technical positions, the probability of this occurring is increased [6].

One example of an effective employee referral program is the *Human Resources Lead Program* (HRLP), which existed when I worked for Sentry Insurance. Sentry was interested in attracting good employees, and the company felt that current employees could serve as a good resource for excellent prospective employees. There were a number of key positions in the organization that Sentry management found to be challenging to staff. To increase the number of qualified candidates for these positions, Sentry implemented the HRLP. Under the program, an employee was paid a cash amount for referring a candidate whom the company hired for one of the key positions. The cash amount varied, depending on the position. The referring employee was paid a portion of the cash amount when the new employee was hired and then received the remainder after the new employee had been with

Sentry for a given period of time. Dividing the cash bonus into two components encouraged the referral of good employees who would be likely to stay with the organization.

Sentry found the HRLP to be effective in providing excellent candidates for certain key positions. One reason why the HRLP generated good candidates was that then-current employees tended to recommend people whom they felt would perform very effectively. The referring employee's reputation could have been affected, either favorably or unfavorably, by the quality of person whom she recommended. As a result, employees tended to suggest primarily top-notch people. If such individuals were hired, the referring employees hoped that this would have a favorable effect on their reputation within the organization.

There are various advantages to both formal and informal employee referral programs. Informal programs obviously are simpler and cheaper to administer. However, formal programs tend to produce better results in general.

If an individual is not hired under a formal program, the referring employee might be more likely to get upset because he would not receive a cash bonus. Under both formal and informal programs, however, a referring employee might occasionally be upset if a person whom he referred is not hired because the employee may feel that he has "let down" the person who was referred. Despite these and perhaps some other minor potential problems, the advantages of an employee referral program greatly outweigh the disadvantages.

1.3.2 Professional Organizations

Another way to attract excellent technical employees involves the use of professional organizations. Obviously, belonging to such a group can greatly enhance a manager's technical expertise and professional development. In addition, a professional organization can be a great source of fine candidates for open technical positions. Some groups have formalized member referral systems that attempt to match up organizational openings with interested candidates [8]. This can be an excellent way to generate some additional outstanding candidates who might fill key positions in an organization.

In addition to the formalized referral programs, however, professional associations can also generate candidates on a more informal basis. For example, a technical manager who works for a really good organization may extol the virtues of the company at a professional meeting. She might informally share with others that, for example, the organization is a great place to work and employees are treated exceptionally well. It is quite possible that such statements may initially or later influence a member of the professional organization to try to find out more about the company. Such an individual may be currently employed but not entirely happy with her current situation. As a result, this person may be very interested in exploring employment opportunities, either now or in the future, with the technical manager.

1.3.3 Networking

Another important technique for recruiting outstanding technical employees is networking. It is important for a manager of technical employees to get to know as many people as possible in other good companies within the industry, as well as in other relevant industries. A manager of technical employees who works for an outstanding company should tell other individuals about his organization. It is quite possible that one of these contacts may at some later point consider making a career change. Knowing that the technical manager works for a great company, the individual might well be interested in finding out more about that organization.

A technical manager can use an adaptation of the approach used by individuals who are involved in the area of executive search. In other words, he can find out where the best people are and get to know them. It is important to do this when a technical manager is recruiting. Preferably, however, a manager will do this before he has a recruiting need. Having these contacts can be extremely valuable. Even if one of the contacts does not want to pursue an opportunity, she might well know someone else who would be interested.

1.3.4 Executive Recruiters

Executive recruiters, as the name implies, recruit executives. However, many also recruit highly paid technical employees as well. Kennedy Publications publishes a book called the *Directory of Executive Recruiters* [9], which can be a useful guide in selecting an appropriate recruiter for a particular organization.

Executive recruiters use a simple premise, that is, identify companies with outstanding reputations in an industry, locate the best employees in these companies, and go after them. Although this strategy may sound simple, it is not necessarily easy to implement, and it is *extremely* time consuming.

There is a great deal of selling involved in executive recruiting. First, the recruiter must persuade a potential client that the quality of candidates that he can locate is worth his fee, namely, about one third of the individual's annual compensation, plus expenses. Obviously, this fee represents a considerable sum. For example, suppose a company is trying to find a top-notch astrophysicist who has had extensive success in the defense industry. If the going rate for such an individual is $100,000.00 a year plus a bonus, the fee would be well over $30,000.00 plus expenses. This may sound like a great deal of money, and it is. However, having done recruiting via the executive recruiting method, I can categorically say that they earn every penny of their money.

In addition to the need to sell potential clients, executive recruiters also must be able to sell candidates on opportunities in other firms. This is often not an easy task. For example, suppose an outstanding organic chemist is working for an excellent company, he loves his job, he is being paid well, and his company provides an exceptional benefits package. It would undoubtedly be quite difficult to convince him to consider interviewing for a position in another organization. However, this is the type of challenge that highly competent executive recruiters successfully meet regularly.

I have had first-hand experience witnessing the persuasiveness of a successful executive recruiter. When I worked for the Parker Pen Company, an executive recruiter from the Chicago area, Jerry Menzel, contacted me about an executive position at Sentry Insurance. When Jerry asked me about the Sentry position, I indicated that I was quite happy at Parker Pen. Despite this, he convinced me to consider the Sentry position because it represented an excellent career advancement opportunity. As it turned out, his persuasiveness paid off for both of us. I was offered and accepted the job at Sentry Insurance.

The process used by executive search firms is very time consuming. First, research must be done to identify the firms that are likely to have candidates for a given position. Next, specific individuals must be identified who could potentially fill the position that is open. In addition to using data bases to locate potential candidates, executive recruiters get many of their names through networking or referrals. Finding someone via the latter two sources is often preferable because it is an initial check on the quality of the candidate. Recommending oneself for a position obviously carries less credibility than being recommended by someone who is well respected in an industry. A great number of hours of library or computer data base research, letter writing, phone calling, phone interviewing, and face-to-face interviewing eventually produces several top-notch candidates for an open position.

Some firms that use an executive recruiter somehow feel that it would be unethical for them to directly recruit employees from their competitors but feel it is acceptable to do so if an executive recruiter is used. This position, from my standpoint, does not make sense. To me, this is similar to believing that if a person hires a "hit man" to murder someone, this is morally acceptable; but if the individual herself kills someone, this would not be morally right. It sometimes surprises me that certain firms feel it is inappropriate for them to recruit from their competitors. Again, this makes no sense to me. Companies fight tooth and nail over customers, and this is considered perfectly acceptable. Why then, is it less acceptable to do battle over top-flight employees?

Actually, as a practical matter, however, it probably is not a good idea to get involved with ongoing internecine warfare with a competitor over prospective employees. However, I see no problem if a company occasionally seeks out an exceptional employee from a competitor. Obviously, the competitor may then try to do the same thing. However, if a company is a good place to work and provides an appropriate compensation and benefits package, the management probably has no great worries regarding significant loss of employees to competitors.

Some companies may feel that they can use the techniques employed by executive recruiters to recruit their own employees, thus saving significant fees. This is a possible strategy. However, I have dealt with the many challenges faced by executives recruiters, and I have experienced first-hand just how time consuming the process is. As a result, I feel that for most companies, using an executive recruiter is an appropriate strategy.

One final note concerning executive recruiters involves how they are paid. True executive recruiters charge a retainer. That is, they charge a fee that is not contingent upon completing a recruiting assignment successfully. The reaction of individuals in certain organizations might be something such as "You mean I have to pay a fee even if we don't get a qualified candidate?" While such an arrangement may initially seem unfair to some people, it is really not unlike the financial arrangements dictated by other professionals such as attorneys and physicians. If someone's doctor fails to cure her of some disease, she still has to pay the doctor's bill. Likewise, if someone's criminal attorney is unable to avoid a fine and/or jail sentence for her client, the client still must pay the attorney's fee. As professionals, executive recruiters expect to be treated similarly.

There may be instances in which a client's actions may influence an executive recruiter's ability to complete an assignment successfully. For example, after a search assignment is begun, a client may change the specifications significantly or a client might decide to stop the search. In such cases, an executive recruiter still needs to be paid for her work up to that point.

1.3.5 Employment Agencies

Employment agencies represent another potential source of qualified technical candidates [7]. Employment agencies differ from executive recruiting firms in that the former are typically paid on a contingency basis. That is, they receive a fee only if they are successful in convincing a firm to hire a candidate whom they have presented. Employment agencies typically charge fees that range from about 25 to 33 percent of the expected annual compensation of the person who is placed. Generally, they tend to recruit candidates at salary levels that are less than those typically dealt with by executive search firms, but there is an overlap between the two. Ordinarily, a recruiter in a contingency-based recruiting firm works with more clients than does her counterpart in a firm that is paid on a retainer basis. This is quite understandable, since a recruiter in a contingency-based firm is paid only if she is successful in placing someone. As a result, she must have quite a few opportunities to receive a fee.

The distinction between retainer-based and contingency-based recruiting firms is not as clear-cut as indicated above, in many cases. For example, there are some executive search firms that charge a partial fee upfront for their services, regardless of the outcome. However, they do not charge their entire normal fee (that is, one-third of a candidate's expected first-year annual compensation) unless someone whom they recommend is actually hired by the client firm.

The fact that a recruiter in a retainer-based firm is paid for his time, regardless of the outcome, does not necessarily mean that he is "better" than another recruiter who works on a contingency basis. As is true with virtually all professionals, there are good, mediocre, and poor recruiters in both retainer-based and contingency-based recruiting firms.

One strategy that can greatly increase the success ratio of an organization using a recruiting firm that operates on a contingency basis is to spend a fair amount of time upfront with the recruiter. A manager of technical employees needs to provide as much information as possible to the recruiter regarding precisely what is needed with regard to the candidates [7]. The more specific a technical manager can be regarding what the candidate specifications are, the more likely it is that the recruiter will understand what is needed and will be able to find a candidate who meets the client's needs.

For example, in one of my previous positions I worked with a contingency-based recruiter to find a technical candidate. I spent considerable time (that is, at least an hour, and probably more) explaining to the recruiter the type of individual needed for the job. The recruiter responded by sending me the resumes of three candidates, all of whom precisely met the specifications I had indicated. One of the three was later hired for the job. There is no question in my mind that had I not invested the initial time with the recruiter, the successful outcome would not have occurred.

Identifying potential contingency-based recruiting firms can be done using a process similar to that used in identifying retainer-based firms. A list of potential firms can be garnered from *The Directory of Executive Recruiters* [9]. In selecting one of the firms on the list to address a recruiting assignment, an organization may evaluate the results of interviews with the potential recruiters. Ideally this would be done in person, but it could also be done on the telephone. In addition, a technical manager could ask the various firms on her list to provide client references regarding the quality of their work. However, perhaps the best way to determine whether a contingency-based firm can help an organization is simply to give the recruiting organization a chance to prove itself on one assignment. If this trial assignment is successful, the organization has developed a valuable resource for future recruiting assignments. If it does not work out well, the technical manager will merely have spent some time that, unfortunately, did not result in a payback. Firms typically spend more time evaluating a retainer-based versus a contingency-based recruiting firm. This is true because if an assignment is not successfully completed by a firm in the former category, a great deal of money as well as time will have been spent on the project.

1.3.6 Indefinite Retention of "Superstar" Resumes

Another approach that can be extremely helpful to a technical manager in the recruiting of outstanding technical individuals is to maintain a separate file of resumes of "superstar" applicants. Virtually every organization gets many unsolicited resumes from various individuals. Probably 98 to 99 percent of the time, there are no openings that match the qualifications of the individuals who send in these unsolicited resumes. As a result, these resumes, if sent directly to technical managers, are typically forwarded to the human resources department for filing and the sending of a "no interest" letter. In most organizations, the human resources department maintains applicant resumes for a period of time in case a relevant opening should arise in the future. I maintained such an applicant file when I was vice president of human resources for SSI Technologies, Inc. I found two major problems with such files, however. The first relates to the retrieval of a resume when it is needed in the future. Some resumes could be filed under a number of headings or job titles. For example, suppose an individual's most recent position was manufacturing engineering manager. Suppose also that in the past, this individual held positions as a product design engineer and a manufacturing engineer. The resume of such an individual could be filed under any of the three aforementioned job titles. The most recent position, manufacturing engineering manager in this case, would probably be the most appropriate job title under which to file this person's resume. However, depending on this individual's situation, he may be more interested in, and perhaps even more suited for, one of the other two positions.

When I recruited for technical positions at SSI Technologies, Inc., there were a number of occasions when I had difficulty locating a resume because the resume was filed under a job title that was different from the one I had expected. In addition, I sometimes was unable to locate the resume of a particular individual that I had remembered seeing at some time in the past because the files had been purged on a regular basis.

One possible solution to the problem of having difficulty locating resumes when they are needed is to use a computerized database. In the case of the example that I used earlier, if the particular system used allowed for retrieval of information under several different job titles, the problem of not being able to locate it could be eliminated. The problem concerning the purging of the files could also be dealt with if the database were not purged, even if the paper resumes were destroyed on a regular basis. However, if an individual whom a manager is trying to locate no longer resides at the address indicated in the database, it would still be difficult to locate that person.

One approach that I recommend for dealing with the two problems of locating and maintaining resumes involves a technical manager making a duplicate copy of any resumes of "superstar" candidates. These resumes could be kept by the technical manager in a special place, even if the human resources department also has a copy of each resume. This provides a good backup, just in case the human resources department has difficulty locating any of these resumes.

I also have a suggestion regarding dealing with the problem of trying to locate someone who no longer lives at the address indicated on her resume. A technical manager can write to all exceptional candidates. In each letter, he could mention that the candidate's qualifications are exceptional. The technical manager could also indicate that he would like to keep track of the individual's future job and address changes, just in case an opportunity at the company may arise. Although not every exceptional candidate will comply with the technical manager's request, many will because they will appreciate the special treatment referred to in the letter.

An example of an exceptional resume that I received while I was vice president of human resources as SSI Technologies was one from a vice president of engineering. Although I never interviewed this person or talked to him on the phone, it was clear from his resume that he was an unusual individual who had accomplished a great deal in his various positions. I kept his resume in a special place so that I could easily locate it if an opening relevant to his background occurred. A position such as the vice president of engineering was particularly critical in a "high-tech" organization such as SSI.

Obviously, setting up a special file for exceptional resumes and following up with individuals on job and address changes takes time. However, the time spent is well worth it because it may enable a technical manager to hire an outstanding technical employee whom he would not have been able to hire, had these special actions not been taken.

1.3.7 Advertising

One of the most commonly used techniques for recruiting technical as well as nontechnical employees is advertising in newspapers or trade journals [3, 6–8]. Some people are very critical of employment advertising. For example, executive recruiters typically say that the best candidates are not looking for a job and, as a result, employment advertising will not reach these individuals. In response to this comment, I would say that it is often true. However, even outstanding individuals sometimes find themselves out of a job. As a result, these exceptional individuals *are* sometimes actively seeking employment, and therefore they could be attracted through employment advertising. Second, depending on where a technical manager places an employment advertisement, even individuals who are not actively seeking employment may see and respond to an advertisement.

Throughout my twenty years of experience, I have recruited quite a number of technical employees and found a number of outstanding candidates through employment advertisements. One particularly important factor that can determine the effectiveness of employment advertising is the labor market. If the labor supply for a particular position greatly exceeds the demands, employment advertising is likely to be a very effective technique. For example, one of my clients, Electrol Specialties Company, needed to recruit a general manager some time ago. An advertisement was placed in the Midwest edition of Tuesday's *Wall Street Journal*. More than 500 resumes were received, many of which were very good. Top management reviewed the 500+ resumes and selected the best. I then phone-screened these individuals, did in-person interviews, and gave selected tests to the best candidates from the telephone interviews. Next, I wrote an in-depth report on the top two candidates, whom I recommended for the job. The top candidate was offered, and accepted, the position.

Employment advertising is not inexpensive. Many technical managers are often surprised at the price of a display ad that is placed in the Sunday edition of a major newspaper. Rates vary greatly, depending on the circulation of the particular newspaper. The greater the circulation, the more expensive the employment advertisement. Since ads are not inexpensive, it is important that an ad have maximum effectiveness and reach the greatest number of people possible. One way to maximize the effectiveness of an ad is to use an organization that specializes in employment advertising [7]. These organizations specialize in writing copy and doing art work, including logos and borders, for example, for employment ads. They can also give good advice regarding which newspapers are best to use and on which days ads are most effective.

If the ad copy is not particularly good, or the artwork is not especially well done, the effectiveness of an employment advertisement is greatly diminished. As a result, it is well worth the investment to get assistance from an employment advertising agency regarding the copy, the artwork, or both. The fees charged by such agencies are quite reasonable, considering the value provided. In fact, many employment advertising agencies have arrangements with certain newspapers that allow them to place ads for a reduced rate. As a result, in some cases an employer can use the services of such an agency and pay the same amount as the organization would have paid for placing an ad directly with a newspaper.

Whether a technical manager writes an ad herself, uses the services of an employment advertising agency, or asks the human resources department to write the ad, it is essential to highlight what a particular organization has to offer an outstanding technical candidate.

For example, one of my clients, GEMPLER'S, Inc., is a catalog direct marketer that has undergone very rapid growth. Working for a fast-growing company would be extremely attractive to many candidates for open positions. As a result, I recommended that the client use this fact when the company ran an employment advertisement for a key position. Some candidates might find the dramatic changes inherent in a very rapidly growing company unappealing. However, such individuals probably would not adjust well to this type of environment. As a result, not having these individuals apply for open positions might be beneficial.

Another way to maximize the effectiveness of employment advertising is to ensure that ads are placed in publications that make it likely that non–job seekers, as well as those who are actively seeking a job, will see and respond to the employment advertisements. For example, placing an employee advertisement in a trade journal [8], preferably in a location that is not exclusively designed for employment ads, is likely to be viewed by those individuals who are not actively seeking employment as well as active job seekers. Likewise, placing an advertisement in a nonadvertising section of a newspaper, such as the sports section, can also serve to attract candidates who are not actively seeking employment as well as those who are [7].

1.3.8 Direct Mail

Another approach that can be very helpful in attracting top-level technical employees is a direct mail campaign [7]. In using this approach, a technical manager might rent the mailing lists of certain relevant professional technical organizations. She might then send a letter to the potential technical employees on these lists.

Using the direct mail approach can be helpful when a technical manager has a specific opening, but it can also help to identify candidates for possible future openings. For example, several years ago, I received a mailing from a well-known organization indicating that the company was seeking psychologists who had a special expertise in employee selection. Although I was not interested in pursuing a position with that organization when I received the mailing, I did keep it on file for possible future reference.

In developing a letter for a direct mail campaign, a technical manager should try to identify the major benefits of working for her organization (for example, the company offers products or services that are on the cutting edge of technology). The letter might also highlight some possible needs of potential candidates that might not currently be met by their present organizations (for example, a technical employee may not be challenged intellectually by his present job). To assist her in designing an extremely effective direct mail letter, a technical manager might enlist the services on an internal or external marketing expert.

Direct mail campaigns do work. Many highly successful organizations sell their products or services via direct mail. Obviously, a technical manager needs to send out a substantial number of letters in a direct mail campaign, as only a small percentage of people who receive a letter actually respond to it. However, if a technical manager is successful in attracting even one or two top-flight potential employees, the efforts are worthwhile.

1.3.9 Universities and Technical Colleges

Universities and technical colleges serve as sources of technical employees for many organizations [7, 8]. However, numerous companies do not necessarily employ the most effective techniques for attracting candidates from these sources.

For example, many organizations maintain contact with universities or technical colleges only when they are actively seeking candidates for open positions (for example, they visit a school only when they are doing on-campus interviews). Organizations need to maintain frequent contact with schools so that the staff members in the placement centers and faculty members in these schools are able to become extremely knowledgeable about their organizations [7, 8]. If a technical manager has repeated contact with, for example, the director of placement in a particular university, the manager can establish a close working relationship with that director. This can be invaluable when students ask the placement director for advice concerning which companies are good places to work.

One way for a technical manager to get additional exposure to both the placement staff as well as students is to visit schools on career days. Such opportunities can be invaluable in terms of promoting the numerous advantages of working for a particular organization. Repeated contacts with schools can have a significant favorable effect upon the number of students who sign up for interviews with a particular organization.

Maintaining a close working relationship with the placement staff of a university or technical school may help a technical manager in identifying not only new graduates but also some of those individuals who graduated in recent years.

1.3.10 Internship Programs

One particularly valuable technique for attracting potential future technical employees is the use of an internship program [8, 10]. Internship programs provide a relatively inexpensive way for an organization to get to know potential future job candidates as well as vice versa. An intern who has a very favorable experience is not only very likely to apply for a position when she graduates, but she may also apply for a position with that organization some years after graduation. Favorably impressed interns also are very likely to make comments to other students, who may also apply for positions at the organizations that are mentioned.

Many organizations currently use internship programs, but some do not use them as effectively as they might. For example, I have talked to some interns who did not have an opportunity to gain an in-depth knowledge of the organization at which they did their internship. I have talked to other interns who did not have a clear idea of what they specifically wanted to get out of the internship experience beforehand or what they actually did get out of the experience afterward.

I have several suggestions that can enhance the effectiveness of internship programs, both for the interns and for the organizations that employ them. These are as follows.

Prior to beginning an internship, a technical manager must specifically define what he hopes an intern will accomplish during the internship period.

A technical manager should request that an intern-to-be identifies specifically what she hopes to accomplish during the internship period. Any discrepancies between an intern's and a technical manager's expectations should be resolved prior to the beginning of the internship.

A technical manager should have frequent contact with an intern during an internship. He should provide frequent encouragement, positive feedback (when appropriate), constructive criticism (when appropriate), and support (when needed).

A technical manager should conduct a thorough performance review with an intern at the end of the internship period. An evaluation of how well the objectives were met should be provided. In addition, a summary of major positive feedback and constructive criticism should be provided. Proper documentation of the performance review discussion should be made for possible future reference.

Interns should be asked to provide an anonymous evaluation of the internship program (obviously, however, if only one person is an intern at a particular time, anonymity will not exist). Specific feedback should be requested concerning favorable aspects of the internship experience, unfavorable aspects, and suggestions for improvement. This information should be used in planning and implementing any changes in the internship program.

1.3.11 Realistic Job Previews

Another technique that can be used to attract top-level technical employees is called a realistic job preview [2]. A realistic job preview differs from what is often presented in customary recruiting. In traditional recruiting, it is assumed that to attract and to "sell" candidates, only the positive aspects of a company and/or a job should be presented. If a technical manager uses the realistic job preview approach, however, she does not just present the positive aspects of a job or an organization. Instead, she highlights what a job is really like, including both favorable as well as unfavorable aspects.

For example, the director of engineering for a corporation may tell a prospective product design engineer that the company's products are the most sophisticated in the industry. However, the director may also add that to develop these cutting-edge products before the competition does, the product design engineers need to put in extremely long hours. He might add that this is especially true just prior to the introduction of a new product. Realistic job previews may have an effect on the selection process. Some candidates, upon learning about some of the unfavorable aspects of a job and/or company, may decide that they do not want to work for the organization. This elimination of candidates via self-selection aids the overall selection process because it tends to eliminate candidates who would not be happy in or do well in a given position.

Some technical managers might feel that using realistic job previews, while improving selection, might hurt the recruiting process, in that their use tends to reduce the number of candidates. However, I believe that realistic job previews enhance the recruiting and selection processes. When a technical manager is totally candid about the unfavorable as well as the favorable aspects of a job or organization, the manager is likely to gain considerable credibility with a prospective candidate. This credibility is an important asset to any organization that is attempting to recruit top-level technical employees. In some cases, it may result in someone accepting a position with an organization rather than taking another position with a different organization. For example, one lab technician told me that she accepted a position with a company because the manager with whom she dealt there was the only one who highlighted the negative as well as the positive aspects of the position and the organization. She felt that this manager was the only one who was truly honest with her. As a result, his company was the one that she selected as an employer.

Some research that I did several years ago reinforces the value of realistic job previews in the recruiting process. While I was with Sentry Insurance, I did a selection test validation study to identify personal characteristics that were correlated with sales success. The research, which was done with then-current sales representatives, correlated their scores on certain personality tests with actual sales performance. The results indicated that the most significant personal variable correlated with sales success was one that might be called self-esteem. The variable that had the second highest correlation with sales performance was one that might be called candor/honesty. In other words, the individuals who were the most forthright tended to be the most effective sales people.

This finding is likely to hold true for products other than insurance. In addition, I believe that it is also true with regard to selling job candidates. Trust is obviously a critical element in any relationship.

1.4 SUMMARY

Several key factors that can help significantly in the recruiting of outstanding employees include providing an excellent compensation package, offering superior employee benefits, and developing a reputation for excellence in employees and in products/services offered.

An outstanding compensation package may include a superior base salary, a generous bonus, and stock or stock options. A superior benefits package serves as evidence that a company truly cares about prospective employees and their families. Developing a reputation for excellence in employees involves having very talented employees who are treated well by the company.

Resources and techniques that can help recruiting include employee referral programs, professional organizations, networking, executive search firms, employment agencies, indefinite maintenance of excellent resumes, employment ads, universities/technical colleges, internships, direct mail campaigns, and realistic job previews.

Employee referral programs are one of the most effective sources of excellent people. Professional organizations often have employee referral services that can be helpful in recruiting. Contacts developed through networking can provide a valuable source of excellent potential employees. Executive search firms can be very helpful in finding talented upper level technical employees. Employment agencies may aid a company in finding some good technical people, and they are generally paid only if someone they refer is hired. Maintaining ongoing contact with outstanding individuals and keeping their resumes on file indefinitely can be of significant benefit to an organization in its recruiting efforts. Employment advertising, one of the most commonly used recruiting methods, can be used to recruit excellent people. Universities and colleges provide a good source of entry-level technical employees. Internship programs can also be a valuable source of outstanding entry-level people. Direct mail recruiting requires sending out many letters, but it can generate some very good candidates. Realistic job previews can aid an organization in the recruiting and selection of outstanding employees.

References

[1] Herzberg, F., B. Mausner, and B. Snyderman, *The Motivation to Work,* New York: Wiley, 1959.

[2] Saal, F., and P. Knight, *Industrial Organizational Psychology: Science and Practice,* Pacific Grove, CA: Brooks/Cole, 1988.

[3] Coss, Frank, *Recruitment Advertising,* New York: American Management Association, 1968.

[4] Egan, Timothy, "Microsoft's Unlikely Millionaires," *New York Times,* June 28, 1992.

[5] Bowes, Lee, *No One Need Apply: Getting and Keeping the Best Workers,* Cambridge, MA: Harvard Business School Press, 1987.

[6] Orlov, Darlene (ed.), *The Hiring Handbook,* Institute for Management, Greenvale, NY, 1986.

[7] Arthur, Diane, *Recruiting, Interviewing, Selecting & Orienting New Employees,* Second Edition, New York: AMACOM, 1991.

[8] Carruth, Donald L., Robert M. Noe, and R. Wayne Mondy, *Staffing the Contemporary Organization: A Guide to Planning, Recruiting, and Selecting for Human Resource Professionals,* New York: Quorum, 1988.

[9] Kennedy, James H. (ed.), *The Directory of Executive Recruiters 1994,* Fitzwilliam, NH: Kennedy Publications, 1994.

[10] Phillips, Jack J., *Recruiting, Training, and Retaining New Employees: Managing the Transition from College to Work,* San Francisco: Jossey-Bass, 1987.

Chapter 2

Selection Standards

2.1 INTRODUCTION

The people who are selected to work in any organization comprise an exceedingly significant element that influences the overall success of the organization. As a result, the standards that an organization uses in the selection process are extremely important.

I agree with a number of other individuals [1–3] who feel that it is essential to set high but realistic selection standards. This means that an organization or a specific technical manager should not be willing to settle for an "adequate" candidate even if he appears to be the best candidate available. Likewise, however, selection standards should not be set at an unrealistically high level. If a technical manager waits until he finds the "perfect" candidate, no one will ever be hired.

I have witnessed situations at both ends of the selection standards continuum. For example, I once conducted a selection assessment of a candidate for a process engineering position. I told the hiring manager that the candidate was likely to be an adequate but not an above average performer. As a result, I suggested that the manager continue to look for a stronger candidate. However, the manager decided to hire the individual anyway. His rationale was that this individual had experience dealing with similar products in other organizations and that this experience was relatively rare. In addition, he was the strongest candidate that had arisen so far in the selection process.

When I talked to the hiring manager about six months later, he indicated that he regretted hiring this individual. The individual had performed adequately, as I had predicted. However, when the manager's department was faced with significant new challenges, the manager had serious doubts as to whether this person would be able to handle them.

Setting unrealistically high selection standards can also cause problems. For example, a manager who supervised a group of technicians for a manufacturer needed to hire an additional technician. The manager wanted someone who had significant expertise in the repair and maintenance of all of the organization's current equipment as well as all of the new equipment that was going to be delivered and installed in the coming months. In addition, the manager expected the individual whom he hired to be willing to work extremely long hours on a regular basis.

A number of qualified technicians were recruited for the position. None of these individuals, however, had significant expertise regarding all of the company's equipment, both the existing and the new equipment that was to be delivered. As a result of the manager's insistence on finding someone who could meet his unrealistic standards, the position was open for nearly nine months. During this period, the existing technicians, who had already been working substantial overtime, had to work every day of the week because the group was one person short. After nine months of working seven days a week, a number of the technicians were beginning to feel a great deal of stress. Problems were arising in their personal lives, and their morale dropped to a very low level. Several of them were on the verge of quitting the organization.

The manager, after becoming aware of the adverse situation in his department, finally agreed to modify his unrealistic selection standards. He agreed to hire a new technician who was well versed concerning most of the equipment and who was willing to receive special training to become competent in the repair of the other equipment. Had the manager simply set more realistic selection standards in the beginning, he could have avoided the significant problems in his department.

2.2 MATCHING THE INDIVIDUAL WITH THE JOB, THE SUPERVISOR, AND THE CULTURE

When a technical manager wants to select a top-flight technical employee, she needs to be concerned about three major areas, that is, the match of each candidate with the job [1, 4, 5], his supervisory preferences/expectations regarding how the job should be done [5], and the organizational culture [1, 3, 5–8]. Regarding the job, the technical manager needs to define very specifically the knowledge, skills, abilities, and personal attributes (KSAPs) that a candidate must possess to perform successfully in the position [1, 4, 8–11]. (See Section A.1 in Appendix A for definitions of these terms.)

For example, suppose a product design manager for a manufacturer of electronic components is interested in hiring a new product design engineer. The manager might feel that a qualified candidate would need to have a basic knowledge of electronics, which would be learned in an undergraduate electrical engineering program. In addition, the manager may believe that it is necessary for a qualified candidate to have knowledge concerning the design of some type of electronic components.

In the manager's opinion, an excellent candidate might need to be skilled in using a computer to design products that can be manufactured easily and inexpensively. In addition, the manager might feel that the ability to manage a product design team effectively is critical. Finally, the manager might believe that having a very strong sense of urgency is a critical personal attribute for any well-qualified candidates. (In most actual situations, a technical manager would probably want to define more KSAPs than were indicated in this example. However, the example provides a guideline for the types of KSAPs that a manager might wish to consider.)

The manager described might be one who manages with relatively loose control. In other words, he may be the type of individual who provides a basic objective to a staff member and then expects her to both determine how to accomplish this objective and achieve the goal without receiving a great deal of guidance or support from the manager. In such a case, it would be important for the manager to define this as a personal preference or expectation; a candidate who needs a great deal of direction from his supervisor probably would not be acceptable to this manager. (Once again, in actual practice, a technical manager would probably want to define more than one supervisory preference.)

The manager in the example also needs to define the type of culture that exists in his department, and in the organization as a whole. This is important because a candidate who does not "fit" the departmental or corporate culture is likely to have some difficulties in performing successfully in the position. For example, suppose that the culture that exists in this manager's department is one in which all employees work closely together and help one another frequently in accomplishing their objectives. If a candidate preferred to work alone on an ongoing basis, she probably would not fit in well with the organization. Thus, such an individual probably would not be a good choice for a product design engineer even if she were qualified in various other respects. (Once again, in an actual situation, a technical manager would typically want to define all of the key aspects of the corporate and/or departmental culture, not just a single one, as in this example.)

In my experience, most managers of technical employees do very little in terms of defining the KSAPs for a position, their supervisory expectations, or the corporate and/or departmental culture. As a result, it is not too surprising that most technical managers do not do a particularly good job of selecting new employees.

2.3 SOME INSTRUMENTS THAT CAN AID THE MATCHING PROCESS

How can a technical manager improve her ability to match candidates for open positions with the KSAPs for the position, her supervisory preferences, and the departmental and/or corporate culture? Some individuals have developed instruments to aid a technical manager in assessing the matches in these areas

[1, 3, 6, 12]. I developed the instruments described in Sections A.2 to A.4 of Appendix A. Section A.2 provides a form that can aid a technical manager in defining the KSAPs for a technical position in her department. Section A.3 describes a tool that can help a manager of technical employees to define her key supervisory preferences/expectations regarding how staff members should do their jobs. The instrument described in Section A.4 can be used to assist a technical manager in defining the key aspects of the culture that exists in her department and/or in the organization as a whole.

By using the information produced in Sections A.2 to A.4 of Appendix A, a technical manager can do an extremely effective job of defining the KSAPs, his supervisory preferences/expectations, and the departmental and/or corporate culture. This information, in turn, can be used to determine which selection devices (for example, interviews, reference checks, and tests) are most appropriate and specifics regarding the selection process (for example, defining interview questions and determining questions that are to be used in reference checking).

The following example illustrates how a manager of technical employees might use Sections A.2 to A.4. John Doe, the manager of manufacturing engineering for the Acme Pen Company (a fictitious organization), has completed Sections A.5 to A.7 with regard to a manufacturing engineer position that he plans to fill in the next month or two. In doing so, he has defined specifically the key KSAPs for the job, his major preferences/expectations as to how the job is to be done, and the key elements of the departmental and corporate culture with which any successful candidate must be able to cope to do the job effectively.

When John compares actual candidates to the various criteria indicated in Sections A.5 to A.7, he probably will find very few people (that is, about 5 percent) who will meet every single criterion. Likewise, he will probably find few candidates (that is, about 5 percent) who do not meet any of the criteria. (This is what I have found in my experience.) As a result, for the majority of the candidates (that is, about 90 percent), he will need to assess whether the match is close enough to consider them for the job. If they do not meet any of the most critical criteria, it probably would be advisable for him to rule them out from further consideration. Of those remaining, the candidates who come closest to meeting all of the various criteria represent the best possible choices for the job.

2.4 SUMMARY

In this chapter, I described how managers of technical employees can develop selection standards regarding positions under their jurisdictions. These standards need to be high, yet realistic. A manager must define specifically the knowledge, skills, abilities, and personal attributes (KSAPs) required to perform the job for which he will be selecting people. The manager also needs to define her preferences/expectations regarding how the job is to be done. Third, the manager needs to define the key aspects of the departmental/organizational culture with which a candidate must be able to effectively cope to succeed in the job. Some forms that a manager can use to define these three areas as well as illustrations of completed forms for a sample technical position are provided in Appendix A.

References

[1] Smart, Bradford D., *Selection Interviewing: A Management Psychologist's Recommended Approach*, New York: John Wiley & Sons, 1983.

[2] Drake, John D., *Interviewing for Managers: Sizing Up People*, New York: AMACOM, 1972.

[3] Mandell, Milton, *The Selection Process*, American Management Association, New York, 1965.

[4] Fear, Richard A., *The Evaluation Interview*, Third Edition, New York: McGraw-Hill, 1984.

[5] Goodale, James, *One to One: Interviewing, Selecting, Appraising, and Counseling Employees*, Englewood Cliffs, NJ: Prentice-Hall, 1992.

[6] Pinsker, Richard J., *Hiring Winners: Profile, Interview, Evaluate: A 3-Step Formula for Success*, New York: AMACOM, 1991.

[7] Buhler, Patricia, "Managing in the 90's: Hiring the Right Person for the Job," *Supervision*, July 1992, pp. 21–23.

[8] Gulliford, Richard, "The Role of Personality in Assessing Management Potential," *Management Decision*, Vol. 30, 1992, pp. 69–75.

[9] Roth, Philip L., and Jeffrey J. McMillan, "The Behavior Description Interview," *CPA J.*, December 1993, pp. 76–79.

[10] Dettore, Albert A., "The Art of the Interview," *Small Business Reports*, February 1992, pp. 11–15.

[11] Levine, Edward L., *Everything You Always Wanted to Know About Job Analysis*, Tampa, FL: Mariner, 1983.

[12] Lopez, Felix M., *Personnel Interviewing: Theory and Practice*, New York: McGraw-Hill, 1975.

Chapter 3

Effective Interviewing

3.1 PLANNING

3.1.1 Introduction

Interview planning refers to the various activities that a technical manager must undertake to prepare for an interview. These activities include reviewing various documents relevant to the candidate and the position and determining what information needs to be gathered during the interview.

3.1.2 Appropriate and Inadequate Planning

What should a technical manager do to plan effectively for an interview? First, he needs to review the job description for the open position to refamiliarize himself with the key tasks that the candidate who is selected will be performing. Next, a technical manager should review the KSAPs [1] (these are defined in Appendix A) that he had identified as being critical, very important, and highly desirable concerning successful job performance. Then, he needs to review his supervisory preferences/expectations that are critical, very important, and highly desirable. Next, a technical manager should review the critical, very important, and highly desirable aspects of the departmental and/or corporate culture with which a successful candidate must be able to deal to perform effectively. He should then review the resumes of the candidates whom he will be interviewing. Finally, he needs to develop interview questions. (This will be explained in the next section.)

A common mistake that many technical managers and many nontechnical managers tend to make is not spending adequate time planning for an interview [1–3]. I have witnessed first-hand numerous examples of poor interview planning.

For example, when I was interviewed by an executive early in my career, he spent about twenty-five minutes extolling the virtues of the company and telling me about his many great accomplishments; only about five minutes was allotted to asking questions about my background. Obviously, an interviewer can learn little or nothing about an interviewee when he is doing all or almost all of the talking. I believe that some technical as well as nontechnical managers fall into this trap (that is, doing too much talking in the interview) because they have not spent adequate time prior to the interview planning questions that they want to ask. As a result, many managers tend to "wing it" and end up rambling on a great deal and learning very little about the candidate. Another example of poor interview planning that I witnessed involved a manager who obviously had not read my resume prior to our interview. This was apparent because he continued to ask questions about information that was clearly indicated on the resume (for example, where I had gone to school and previous jobs I had held). Using valuable interview time simply to rehash what is already included in a resume is not only a waste of time for both the interviewer and interviewee but also a subtle insult to the interviewee. It is an insult because the manager obviously has not bothered to take the time to read the candidate's resume thoroughly before the interview session. Such behavior may communicate to some candidates that the cover letter and resume that she spent time writing were not viewed as important enough for the manager to spend his time reviewing.

3.1.3 Developing Interview Questions

A technical manager should identify potential interview questions that measure a candidate's match with the job, the supervisory preferences/expectations, and the culture. It is very helpful to have a standard list of questions that can make up a type of "question bank" or "question pool" [4, 5]. The manager can then select the appropriate questions from the pool of questions that are relevant to the specific open position. A list of some sample questions that could be used in a question pool is indicated in Section B.1 of Appendix B.

In addition to the standard questions that are taken from the bank, a technical manager may need to add some customized questions. One type of customized question might arise based on a technical manager's review of a candidate's resume. For example, a candidate's resume may have a time gap indicated between two previous positions. An example of a customized question that addresses this issue is, "What did you do between the time that you left employer C and started working for employer D?"

A technical manager may also need to develop customized questions when there are key KSAPs, supervisory preferences/expectations, or aspects of the departmental and/or corporate culture that are not addressed by the standard questions in the question bank. For example, a director of manufacturing engineering may be in the process of selecting a new manufacturing engineer. One of the key abilities relevant to this job might be the ability to "juggle many balls simultaneously." If the director of manufacturing engineering feels that there are no standard questions that measure this ability, she may need to develop a customized question to address this issue.

One type of customized question that can be used in situations such as this might be called the "behavioral illustration" type question [6]. This type of question involves asking an individual to indicate examples of his behavior from the past that illustrate that he meets a certain criterion for a position. Using the previous example involving the director of manufacturing engineering who is seeking a manufacturing engineer who can juggle numerous balls simultaneously, an example of a behavioral illustration type question would be, "Describe a situation where you needed to complete three or more tasks at the same time." Depending on the response given to this question, follow-up questions could also be asked to determine whether any tasks were not completed and to ascertain how he felt about the situation.

Suppose a technical manager has developed a list of interview questions that cover all of the KSAPs, her supervisory preferences/expectations, and the key aspects of the departmental and/or corporate culture. Suppose also that the manager has carefully reviewed the applicant's resume and has developed some additional questions that deal with a few concerns and/or discrepancies noted on the resume. At this point, the technical manager is ready to schedule an interview with the applicant. When setting a time for an interview with an applicant, one of the most common mistakes that technical managers make is to set aside insufficient time for an adequate interview. Many technical managers set aside thirty to sixty minutes for each applicant interview, and they feel that this is plenty of time. Unfortunately, this is definitely not the case.

When I questioned certain technical managers about whether they have set aside adequate time for an interview, a number of people indicated that they could not think of anything to talk about for more than forty-five to sixty minutes. However, when a technical manager has prepared in the manner described, he will have enough questions to generate discussion for at least an hour to an hour and a half, and perhaps even longer. In addition, time must be set aside to answer an applicant's questions about the job, the department, and the company; to provide information about these three areas in addition to what an applicant asks about; and perhaps to give the applicant a tour of the facility. Thus, the total time that typically should be scheduled is probably somewhere in the range of two to three hours. If it is not practical to do this in one session, the candidate could be brought back a second time.

3.1.4 Legal Issues In Interviewing

Most technical managers are aware of the fact that there are some interview questions that should be avoided for legal reasons. Some of the questions to be avoided are rather obvious. For example, a manager should never ask a candidate, "How old are you?" However, some other questions that should be avoided are much less obvious. In addition, the laws that are relevant to selection interviewing do change from time to time.

Section B.2 in Appendix B provides a short quiz that a technical manager can use to assess his current knowledge of legal issues related to selection interviewing. Section B.3 provides information that a technical manager can use to evaluate his responses to the quiz in Section B.2 (that is, it gives the technical manager an idea as to whether each of the questions posed in Section B.2 does or does not have any associated potential legal problems).

For a more complete coverage of legal issues involved in interviewing, a technical manager should find a good book that deals with this topic, such as the one written by Kahn et al. [7] or Berry et al. [8].

3.2 CONDUCTING THE INTERVIEW

3.2.1 Some Key Considerations

As indicated in the previous section, if a technical manager has adequately prepared for an interview and has allotted two to three hours for the process, the manager should be ready to conduct the interview session. There are a number of key points that are important concerning conducting the interview session. While these may seem quite obvious to many technical managers, apparently they are not obvious to others, because I have seen numerous managers ignore these suggestions.

First, it is essential that a technical manager conduct the interview in an environment that is free from interruptions (for example, telephone calls and drop-in visitors). Conducting an effective interview is a challenging process, and it demands a technical manager's full attention. If a manager is interrupted several times during an interview by phone calls, for example, it breaks his train of thought. In addition, by taking several phone calls during an interview, a manager may communicate an unintended message to a candidate, namely, that she is not important enough to merit the manager's complete attention. Some technical candidates may assume, correctly or incorrectly, that if they are unable to command a manager's complete attention while he is trying to recruit them, it may be unlikely that they will be able to maintain the manager's complete attention after they are hired (for example, when they need help on a particular project).

A second important consideration to remember concerning interviewing is to maintain privacy. A manager may well be covering various issues during the session that a candidate would prefer to be kept confidential (for example, salary and current job). If other people are within earshot of the conversation, a candidate may feel somewhat uncomfortable. Although some technical managers may have an office that is not completely private (for example, the walls of the office do not reach the ceiling and others outside the office may, therefore, be able to hear some of what is said), a manager can ordinarily reserve a private conference room to conduct the interview.

It is also important that a manager ensure that the furniture in the interview room is comfortable. Having to sit on a hard, uncomfortable chair for several hours will not help to relax a candidate; a candidate who is uncomfortable physically may possibly be psychologically less comfortable being extremely candid during the interview session. Obviously, it is important that a manager do whatever she can to get a candidate to open up during the session. Attempting to create a nonthreatening, relaxed atmosphere can aid in eliciting candid responses from a candidate. For example, a technical manager might tend to spend a few minutes at the beginning of a session with some brief discussion about a candidate's trip, for example. This ordinarily helps a candidate to relax and to get the feeling that he is about to engage in a friendly conversation, as opposed to an inquisition.

Another important issue of which all technical managers should be aware prior to doing selection interviewing concerns how much time a manager should spend talking versus listening. Many inexperienced interviewers tend to do too much talking. One of the reasons that this is true is that such individuals often run out of questions after a relatively short period of time, since they typically have not spent much time planning for the interview. As a result, they typically spend an inordinate amount of time telling a candidate about the company or themselves and not enough time trying to ascertain whether the candidate is a good match with the job, themselves, and the organization.

What percentages of time are ideal regarding listening versus talking in an interview? Some professionals advocate a 90/10 split between the two activities [4]. I feel that ideally the interviewer should be spending about 80 percent of her time listening and only about 20 percent of her time talking during an interview. Why are these percentages ideal? Obviously, if an interviewer spends more than 20 percent of her time talking in an interview, she is limited regarding how much can be learned about the candidate. Conversely, if the interviewer spends less than 20 percent of her time talking during an interview session, it may mean that she is not maintaining control of the discussion. For example, I interviewed one candidate for a technical position a number of years ago, and he clearly had his own agenda regarding what he wanted to cover during the interview. As a result, I asked him one question and he proceeded to answer in significantly more detail than was necessary or desirable from my standpoint. Then he made an attempt to make a transition to his own agenda in terms of covering what he wanted during the interview session. After a relatively short time, I cut off his comments and redirected the interview in the direction in which I wanted it to go.

There are many instances, however, when it is appropriate and even desirable to let a candidate talk for a considerable amount of time without interruption. Basically, I do this when a person answers a question that I have asked and then proceeds to answer some additional questions that I have not yet asked but plan to ask later. I have had several interviews in the past when the technical candidate has continued to talk for as long as ten minutes without my interruption. I chose not to interrupt because I was learning a great deal about the candidate and he was simply anticipating additional questions that I planned to ask.

For example, I interviewed one candidate for a manufacturing engineering position. I asked him what he enjoyed about that position. He proceeded to answer that question. In addition, he told me what some of his key accomplishments were, what some of his dislikes concerning the position were, and why he left the job. Since I had planned to ask him questions about these other areas, I simply let him continue to talk after he answered my question.

The interviewer needs to demonstrate appropriate control in an interview [4, 5], meaning that he must use a structured format regarding questions to be asked. This format is then followed, but not necessarily in a precise manner. It is important to maintain control of the interview, but the interviewer does not want to convey to the candidate that she has no control whatsoever in terms of the direction in which the interview goes.

One of the most frequent problems I have encountered in conducting interviews with technical as well as nontechnical potential employees is that many people become verbose. I typically do not cut a person off the first time this happens because I want to maintain good rapport with the candidate and because she may well say something quite significant. Often, a candidate may simply add an aside that is quite significant and that typically leads me to ask follow-up questions. For example, I have asked many technical candidates about their grade point average in college. Many have indicated this, but then have added that their participation in many activities had some effect on their grades. This has typically led to my asking questions about the nature of the activities in which the candidate was involved. A follow-up question concerning activities can tell a great deal about a candidate in terms of her interests and abilities. When I interview technical candidates who have been out of school for quite some time, I still ask about college and high school grade point averages. However, I often do not ask about activities, particularly in high school, unless the candidate specifically mentions them.

Typically, technical managers should be asking primarily "open-ended" versus "closed" questions [2]. An open-ended question is simply one that cannot be answered with a "yes" or "no" or other one- or two-word response. For example, one question that might be asked is, "How were your grades in college?" Many candidates answer this question in several sentences and with sufficient specificity. However, some candidates tend to give a very vague response, such as, "pretty good." When this occurs, it is a good idea to use a closed question to zero in on a specific response. For example, a technical manager might follow up with a question such as, "What was your rank in class?"

Although most of the questions asked should be of the open-ended variety, as indicated previously, closed questions can serve a very useful function [2]. In addition to forcing a candidate to be much more specific if he has provided a rather vague response previously, closed questions can also be used to clarify a situation. For example, I have interviewed some technical candidates for open positions who have provided rather long-winded explanations as to why they left a particular position. Sometimes, even after a rather verbose explanation, it is still not clear to me whether the person left voluntarily or involuntarily. In such cases, I typically ask a very direct closed question such as, "Were you let go from that position?"

Another key issue concerning conducting an interview is the subject of note taking. There are some individuals who feel that taking notes interferes with the development of effective rapport in an interview. These people believe that an interviewer cannot successfully pay attention to what a person is saying, note her body language, maintain positive rapport, and take notes all at the same time. I believe, however, that although such a task may be difficult, it can be done. Good note taking can be developed through effective practice, just as any other ability. It is important for an interviewer not to concentrate entirely on taking notes. By being overly focused on note taking, a great deal is lost in terms of body language. In addition, some aspects of the rapport that develops between the interviewer and interviewee may be jeopardized.

On the other hand, with practice, an interviewer can learn to take notes and still pay attention to the content of what is said, the body language, and the rapport that is developed. With enough practice at note taking, this task becomes a habit, and doing it becomes second nature to an interviewer. An individual who has developed her note-taking skill appreciably is able to maintain positive rapport by regularly looking up to reassume eye contact [4].

As a psychologist, I am a firm believer in detailed, thorough notes. These should include, as much as possible, the content of what was said, as well as any observations concerning body language, for example. For instance, if an interviewee is asked about his reason for leaving the previous job and he immediately blushes, swallows hard, and begins to breathe more rapidly, these observations should be noted. If the interviewee then says that he simply left the position as a result of getting a better career opportunity, the content of this message would not appear to be consistent with the previously indicated body language. As a result, the interviewer in this case might want to probe further to determine if there were additional, more significant reasons as to why the interviewee left his last job.

One common mistake that novice interviewers make is jotting down only particularly noteworthy candidate comments. For example, a candidate may talk for ten minutes or longer without the interviewer taking any notes. Then, however, immediately after the interviewee says that she did not get along with the previous supervisor because the individual was too detail-oriented, the interviewer might jot something down on a pad of paper. In such a case, it is likely that the interviewee would suspect that she said something that perhaps she should not have. As a result, she might be on guard and less candid during the remainder of the session. On the other hand, if an interviewer is taking notes consistently throughout a session, these significant comments, particularly the negative ones, will not be obvious to the interviewee. In addition, the interviewer will have a much more complete record of what is said. This is particularly advantageous if he does not have a chance to go over the notes for a day or two. However, even if the interviewer reviews his notes shortly after the interview, having significantly detailed notes will enable him to do a much more effective job of interpreting the information than if only brief, sketchy notes exist [4].

3.2.2 Rapport-Enhancing Techniques

It is important for an interviewer to establish good rapport with an interviewee. The quality of rapport reflects the extent to which a candidate feels comfortable and is candid with the interviewer. Having effective rapport is important because it aids an interviewer in obtaining the information she needs regarding a candidate.

A technical manager who is conducting interviews can increase her effectiveness considerably by using certain rapport-enhancing techniques. Some of these will be discussed in the paragraphs that follow. If a technical manager is interested in finding out more about these and other techniques for building rapport, she might review a good book on interviewing, such as one by Smart or Fear [4, 5].

One such technique involves the intentional use of silence [4]. This technique helps a technical manager draw out spontaneous information. It also may encourage a candidate to elaborate on a previous point or to make additional points, and it allows the technical manager to do less talking.

To use intentional silence properly, a technical manager needs to be certain that she maintains eye contact [5]. If a manager is looking down at her note pad, a candidate often will assume that she is formulating a new question, and the candidate will typically wait for the new question to be asked. Naturally, if a candidate does not respond after a reasonable amount of time, a technical manager needs to ask another question or ask for elaboration on the previous point. However, in most cases, simply allowing for a long enough period of intentional silence will encourage the candidate to continue.

Another technique that can help to build rapport in an interview involves the use of a few brief words such as, "I see" [4]. This particular technique can be helpful in that it lets a candidate know that a technical manager cares about what is being said and wants him to continue.

Another rapport-enhancing technique involves repeating a few of a candidate's key words [4]. This technique is helpful in that it encourages a candidate to explain something in more detail. For example, a candidate may make a statement such as "Well, I only stayed in that position for about three months because my immediate supervisor and I just didn't see eye to eye." To use this technique, a technical manager might simply say something such as "Didn't see eye to eye?"

Another technique that a technical manager can use to help build rapport with an interviewee involves using positive feedback [5]. This technique can be helpful in that it often encourages candidates to share additional favorable information. In addition, using positive feedback helps the candidate realize that the interviewer appreciates the value of his accomplishments, and as a result, the candidate may be more likely to share potentially negative information with the interviewer later in the session [5].

The following example illustrates the effective use of positive feedback as a rapport-enhancing technique. A candidate for an open position may say something such as "While I worked in that position, I developed several pressure sensors that were key elements of some of the products we made for the medical market. These products that contained the pressure sensors I designed generated about 30 million dollars a year." In response to this comment, an interviewer might say, "That's really great!" or "You must be really proud of that accomplishment." When he hears such a comment from an interviewer, a candidate typically recognizes that the interviewer values his accomplishments, and the rapport between the candidate and the interviewer is likely to be strengthened.

Another technique that an interviewer can use to help increase rapport with a candidate involves paraphrasing something that a candidate has said [4]. If an interviewer has accurately paraphrased one of the candidate's statements, he tends to feel that the interviewer has been listening carefully. If, on the other hand, an interviewer occasionally does not accurately paraphrase what a candidate has said, this technique allows for a correction of the interviewer's misconception. This technique can help to build rapport because it demonstrates that an interviewer values a candidate's comments; this is indicated by an interviewer's ability to reflect back to the candidate what was said.

The following example illustrates the use of paraphrasing as a rapport-building technique:

> *Candidate:* "I quit that job because I was gone so much that I never got a chance to see my family, and the company sales were just flat."
>
> *Interviewer:* "So you're saying that you left the position because the travel was too extensive and because the company was not growing."

Another rapport-enhancing technique that an interviewer can use involves trying to make questions a bit less blunt [5]. By using this technique, an interviewer may tend to encourage a candidate to be less reluctant to volunteer potentially negative information. An example of the use of this technique would be, "What were some of the factors that may have influenced you in your decision to leave that organization?" It is quite obvious that the aforementioned question is much softer than a question such as, "Why did you quit?"

One final rapport-enhancing technique that an interviewer can use to help strengthen his relationship with a candidate involves minimizing negative data [5]. By using this technique, an interviewer tries to convey to an interviewee that the information she has just shared is not as adverse as the candidate may think. For example, a candidate may have just indicated to an interviewer that she feels that she made a big mistake by quitting a previous job before getting another one. In response to this statement an interviewer may say "Well, we all make mistakes. The fact that we are able to recognize them is an indication that we are learning from them, and as a result, we are much less likely to make the same mistakes again in the future."

3.2.3 Control Techniques

Previously, I mentioned that it is important that an interviewer maintain appropriate control in an interview. Although an interviewee should be given some latitude so that he feels that he has an opportunity to have some control over what happens in an interview, the interviewer has to be the individual who has primary control of the session.

One reason why an interviewer needs to control an interview is that it is important that she ensures that an interviewee does not avoid or gloss over important topics [5]. For example, I have conducted interviews in the past in which candidates have attempted to avoid talking about their reasons for leaving certain jobs. Sometimes an interviewee says something such as, "So then I left that position and started working for the XYZ Corporation. That was a really interesting position. Let me tell you about that." In such a case, it is important for the interviewer to interrupt the interviewee and indicate that before talking about the XYZ Corporation, which is a topic that she would like to discuss, the interviewer would like to review some of the reasons that might have influenced the interviewee's decision to leave the previous company.

Another reason why an interviewer needs to maintain control in an interview is that it is important that she ensure that a candidate does not provide extensive, unnecessary detail early in the session; otherwise, the interviewer may not have sufficient time to complete the interview [5]. If readers wish to learn more about the topic of interview control, they are referred to the books by Smart and Fear [4, 5].

There are a number of interview control techniques that can be used when an interviewee resists providing negative information. One such technique involves simply asking the question again [4]. For example, suppose an interviewer had asked an interviewee a multiple-part question concerning his perceived strengths and developmental needs regarding the position and the interviewee only indicated his strengths. In this case, the interviewer may need to repeat the second part of the question again.

Another approach involves requesting at least some minimal information that might be construed as being unfavorable [4]. For example, an interviewer might ask an interviewee about criticism she received in her current position from the supervisor. If the candidate indicates that she has received no major criticism, the interviewer might ask a question such as, "Even though you may not have received any *major* criticism, surely there must be at least a few *minor* areas targeted by your supervisor for improvement."

An additional technique that can be used to elicit unfavorable information from a candidate involves the amplification of favorable information [4]. For example, suppose an interviewer asks about a candidate's accomplishments with regard to a former position. After the candidate has mentioned these accomplishments, the interviewer might ask the candidate to elaborate on each accomplishment in some detail. After having done so, most candidates will be much more receptive to sharing information that might be construed as somewhat unfavorable.

Self-assessment questions can be very helpful with regard to eliciting both favorable and unfavorable information from candidates [4]. An example of such a question is, "What do you see as your main assets or strengths regarding this position, and what do you feel are some of your developmental needs regarding this job?"

Since self-assessment questions such as that just mentioned ask candidates for both favorable and unfavorable information, many candidates are willing to share at least some negative information about themselves. However, some candidates may be unwilling to share even minimal unfavorable information. As a result, it may be necessary for an interviewer to use one or more techniques to elicit unfavorable information from a reluctant candidate.

One approach to aid in eliciting unfavorable information is to ensure that a candidate is not interrupted as he mentions negatives [4]. When a candidate mentions some unfavorable information about himself and an interviewer immediately asks the candidate to explain this in some detail, an interviewee will often begin to "clam up." A much more effective approach is to remain silent until a candidate has mentioned quite a number of areas needing improvement. When it is clear that a candidate is finished with this list, an interviewer can then ask him to explain any of the negative information that warrants clarification.

Another approach that an interviewer can use to deal with an interviewee who is reluctant to share developmental needs in response to a self-assessment question is to ask the interviewee to take some additional time to think about the question [4]. This technique may sometimes elicit additional unfavorable information concerning the candidate.

One final technique [4] that an interviewer might use to elicit unfavorable information from a reluctant candidate is to make a statement such as, "You've done a very good job of mentioning a number of assets that you think you could bring to this position. In addition, I am very interested in your assessment of some possible areas needing development regarding this job. No one is perfect, and each of us can improve in certain ways as part of our overall growth process."

3.3 CONCLUDING THE INTERVIEW

A self-assessment question similar to the one indicated in the previous section is often an ideal question to use when nearing the end of an interview. In addition, however, an interviewer might also ask a candidate if she has anything else to add that was not discussed previously. Next, it is also a very good idea to provide a candidate with more information about the position, the department, and the organization [9]. An interviewer should also spend some time "selling" a candidate on the job and the company [9]. I feel that this is a good idea even if the interviewer has made a preliminary assessment that indicates that the candidate is not a good match. Even if the candidate is not the right person for the job, she may know other individuals who might be a good fit for the position. Alternatively, the candidate herself may be suitable for a different position. Also, if a candidate is treated well and has a very favorable impression of the interviewer as well as the organization, it is quite possible that she may mention something favorable about the organization in the future that may influence another potential candidate to contact the organization. Finally, many candidates are customers of the company, potential customers, or may know customers or potential customers. A favorable impression created during an interview may help to gain and/or retain customers.

After an interviewer has had a chance to tell a candidate about the job, the department, and the organization and to sell the candidate, the interviewer should ask the candidate if she has any questions about the job or the organization [9].

After this, a candidate should be given some sort of idea regarding the time frame within which she will receive feedback regarding a selection decision [9]. This is important because many candidates are either reluctant or forget to ask a question like this. Many candidates, when they are not given any idea concerning the next contact, may assume that they are no longer viable candidates, even when this is not the case.

For example, when I was with a previous employer, one candidate was interviewed by several company employees, one of whom left the country immediately after his interview. Since this individual did not have a chance to give anyone in the company any feedback regarding his interview impressions, a decision regarding this candidate was postponed for several weeks until he returned from overseas.

The candidate in this case, who was in fact later selected for the job, assumed that he was not going to get the position because he had not heard from the company within two weeks. Fortunately, before he was ready to accept an offer from another organization, he called the company just to make sure that they were, in fact, no longer interested in him. Had he not done so, the company probably would have lost a candidate whom they wanted to hire.

The final task that an interviewer should be sure to complete at the end of the session involves thanking a candidate for her time [9]. Doing this, besides involving common courtesy, may result in a number of potential benefits to the organization that were referred to previously in the section on selling the candidate.

3.4 INTERPRETATION OF INTERVIEW INFORMATION

3.4.1 Intuitive Versus Analytical Approach

This section addresses some very specific information concerning the analysis and interpretation of interview data. However, before beginning the discussion of specific psychological principles of observation and interpretation, consider some general advice regarding the interpretation of interviews. Most people seem to be primarily either "left-brain" or "right-brain" thinkers; that is, most individuals seem to favor either analytical (left brain) or intuitive (right brain) processes in their thinking.

Many engineers and other technical employees, as well as many technical managers, tend to be primarily analytical rather than intuitive thinkers. My favorite approach, and probably my primary approach, to interviewing is also a very analytical one. However, I also recognize, and I feel that it is important for technical managers to be aware, that an intuitive approach to interviewing is potentially as valuable as one that focuses more on analysis.

An intuitive approach to interviewing, because it involves the use of a different part of the brain from a more analytical approach, tends to generate different information. This additional, different information can be very valuable in an interview. For example, the vice president of technology for a high-technology manufacturer may intuitively pick up some negative information about a candidate for a product design engineer position during an interview. Although the vice president may not know why he feels uncomfortable about the candidate, it is important that he not ignore these feelings generated by his intuition. Instead, it is important that the vice president try to determine the reason for his discomfort. Sometimes, this may become apparent after the person has had a chance to combine his analysis with intuition.

For example, the vice president in this example may go back to the interview notes and recognize that he is uncomfortable about the fact that the candidate was in a particular fraternity in college that had rejected the interviewer in his attempt to become a member. In this case, the interviewer's feeling of discomfort would be based on an inappropriate bias and should be ignored.

On the other hand, this same vice president of technology may go back to his interview notes and determine that he feels uncomfortable with the candidate because she had reported engaging in certain behavior that the interviewer feels is not entirely ethical. In such a case, the interviewer's intuitive feeling of discomfort may well be appropriate. If an interviewer has some questions about a candidate's moral standards and some doubts as to whether the candidate can be trusted, it is unlikely that the interviewee would be a good fit for the position.

Although, as I indicated previously, most people tend to be primarily intuitive or analytical, I believe that anyone can learn to develop their thinking using the nondominant hemisphere of the brain. Learning to make use of information gained from both hemispheres of the brain can enable a technical manager to become an even more effective interviewer.

3.4.2 Some Preliminary Suggestions

In the remainder of this section, I offer a number of preliminary suggestions regarding interview interpretation, discuss the interpretation of work experience and educational background, and make some concluding points. If readers want to find out more about some of these and other points, they might consult books by Smart and Fear [4, 5].

Technical managers need to develop initial impressions [5] during an interview concerning how well a candidate matches the job, the supervisory preferences/expectations, and the organizational culture. These initial impressions need to remain tentative until a technical manager has sufficient evidence to justify more firm conclusions.

A technical manager should always look for more than one piece of information concerning each initial impression [5]. It is important for managers to keep an open mind. They need to look for additional data that may either confirm or negate their initial impressions regarding a candidate [5]. Technical managers must keep in mind the fact that these impressions regarding candidates need to focus on the key aspects of the job, their supervisory preferences/expectations, and the organizational culture.

Another key point that technical managers need to keep in mind is that they should be wary of extremes noted in candidates [4]. Some of these extreme characteristics may reflect not only possible assets but also potential drawbacks in certain situations. It is not necessarily true that *every* strength that a candidate demonstrates can serve as a weakness as well. However, there are many cases in which this is true. For example, a candidate may appear to be extremely direct and forceful during an interview. Although assertiveness is generally considered to be a favorable characteristic, some individuals go beyond assertiveness. In some cases, such individuals may become blunt and tactless.

Another example of how a strength may operate as a "double-edged sword" is illustrated by an individual who is extremely thorough and detail oriented. Thoroughness is generally considered to be an asset in most situations. However, some individuals carry thoroughness to the point of being compulsive and, consequently, get bogged down in details.

Another key point that is important for an interviewer to keep in mind is that the best indicator of what a candidate is likely to do in the future is typically past performance [4]. For example, if a technical manager is interviewing a candidate who has had numerous jobs since college, none for more than a relatively short period of time, then it is quite likely that this person would not be employed for a very long period of time at the manager's company.

3.4.3 Interpretation of Work Experience

One important topic about which a technical manager should always ask interviewees is the results achieved in various employment positions [4]. Both the quantity and the quality of what candidates have accomplished in a given period of time can be very useful information in forming comparisons.

When candidates discuss their accomplishments, they often reveal a sense of urgency or lack thereof. In addition, this topic may also reveal whether their need for achievement is extremely strong, moderate, or below average. Finally, discussing this topic often will provide some evidence concerning a person's work ethic. A candidate may provide some information that suggests that her work ethic is considerable, average, or below average.

Technical managers should note whether a candidate's accomplishments are brought forth easily, with little prompting, or not. Interviewees who have an extremely high need for achievement typically have no problem whatsoever in noting a number of significant specific accomplishments regarding each job that they have held in the past. Such individuals, in a sense, speak the "language of achievement." On the other hand, individuals who have only a moderate or limited need for achievement often have some difficulty in coming up with accomplishments in one or more of their positions.

Another topic that can reveal a great deal about a candidate in comparison to others is what she liked or disliked about each position [5]. For example, a candidate may indicate that she did not like a particular position because there was too much routine involved. If the position for which the candidate is interviewing currently has very few routine activities, this would not be a potential problem. However, if the position does have quite a few rather mundane activities, this candidate may not be well suited for the job.

What an individual says about his likes in various positions may reveal information about possible assets [5]. For example, a candidate who says that he enjoys doing mathematical calculations is probably good at tasks requiring mathematical ability. On the other hand, what a person has enjoyed about various positions may also reveal potential developmental needs [5]. For example, a candidate who enjoyed the fact that a job had very predictable hours is probably not an ideal candidate for a position that is likely to require a great deal of sporadic overtime.

Similarly, dislikes can also reveal potential strengths and developmental needs [5]. For example, if a person disliked the lack of challenge in a previous job, it is possible that this is one indication that the person has a relatively high need for achievement. On the other hand, suppose a person mentions a dislike of the lack of structure in a previous job. It is possible that this person is not particularly strong regarding planning and organizational skills and/or may lack the initiative needed to provide his own structure in a position.

Another topic that is worth probing in an interview concerns compensation [5]. Although it is far from a perfect barometer of competence, salary progress can be one indicator of an individual's capabilities. Naturally, other factors also need to be taken into account, such as various companies' philosophies on pay. For example, a candidate may report that his last few merit increases at an organization averaged about five percent. If an interviewer finds out that the average merit increase given in that organization is about six percent, this obviously reveals something different about the candidate than if the average increase for the organization were only two percent.

One aspect of compensation of which an interviewer should be cautious concerns the candidate who has shown good salary progress between companies but only limited progress within companies. Such an individual may be someone who interviews very well but who lacks substance and does not deliver in terms of performance [4].

Another topic about which it is important to ask concerns the reasons why a candidate left previous organizations or is considering leaving the current organization if he is still employed [5]. A candidate's responses concerning reasons for leaving various positions often may provide the interviewer with information that is relevant to how well he matches the job, supervisory expectations, and organizational culture of the interviewer's company. For example, a technical manager may feel that it is important for a candidate to be capable of functioning somewhat autonomously and yet willing to take direction. If the manager interviews a candidate who indicates that he has left several positions because the supervisor did not provide him with enough latitude, this may be a potential indicator of the interviewee's possible unwillingness to take direction. Asking this candidate additional questions about how much autonomy he requires and about the amount of direction he is willing to accept can provide further important information for the technical manager concerning this candidate.

One important point that a technical manager should remember is that she should never accept "I had a better opportunity" as an adequate explanation for leaving a previous job [4]. While a candidate often may leave one position because another provided a better opportunity, this may sometimes simply be a cover-up for some of the more significant reasons for leaving. A technical manager should always probe to find out *all* of the various reasons why a candidate left a particular position [4]. People seldom do anything for one single reason. There is nearly always more than one reason for any behavior, other than extremely simple behavior, including leaving an organization.

The following four points summarize some of the key tenets of interpretation of interview information that have been discussed thus far.

1. Initial impressions concerning how well a candidate matches the job, the supervisory preferences/expectations, and the organizational culture should be developed. The interviewer needs to seek out additional information that can serve either to confirm or to negate these initial impressions.

2. Be wary of extremes. They can sometimes be double-edged swords (that is, extreme characteristics can sometimes be harmful as well as helpful).
3. The best single predictor of how a candidate will perform in the future is probably past performance.
4. People seldom do anything for one single reason (with the possible exception of the performance of very basic activities). As a result, interviewers need to probe for additional motives when a candidate provides a single one (for example, a candidate says that she left a job "for a better opportunity").

Although these tenets are somewhat fundamental, they are extremely important. My own observations of the interviewing done by various technical managers suggest that many of them are unaware of one or more of these rather basic ideas. As a result, it is critical that all technical managers keep these in mind as they conduct interviews.

One rather simple but rather effective way of keeping the aforementioned tenets in mind is to jot them down on a 3 × 5 card, which a technical manager could keep in his desk. Prior to conducting any employment interviews, a manager could simply take out the card and review the various tenets immediately prior to doing an interview.

3.4.4 Interpretation of Educational Background

Although the major portion of an interview should be devoted to the discussion of previous work experience, it is important for an interviewer to discuss the candidate's educational background as well. Obviously, candidates who have little or no previous work experience cannot spend much time talking about work. As a result, an interviewer needs to focus a great deal of time on discussing the educational background of such candidates. However, it is a good idea for an interviewer to spend at least some time discussing educational background with every candidate, even those who have been out of school for quite some time and who have a great deal of work experience.

What is the value of discussing a candidate's educational background? It can give the interviewer information concerning the candidate's interests, personal characteristics, and abilities and even some potential information concerning the candidate's overall intellectual functioning [5].

One area that can be covered in the educational background section concerns subjects that a candidate liked and disliked. Although the correlation is far from perfect, interests are often related to abilities [5]. For example, if a candidate for a technical position indicated that she greatly enjoyed the courses taken in mathematics and science, it is fairly likely that the candidate did well in these courses and has rather good analytical skills. Enjoying and doing well in subjects in mathematics and science will be quite common for candidates for technical positions. In fact, if a candidate indicated that he did not enjoy and/or do well in subjects in math or science, the interviewer should certainly pursue this matter further.

If a candidate indicated that she greatly enjoyed subjects requiring verbal ability, such as foreign languages and English, it is quite likely that the candidate has rather good verbal skills. Having these skills would be quite valuable in a technical position involving a great deal of communication, either written or oral. An example would be a product design engineer who needs to make presentations to prospective and/or current customers. Such an individual certainly would need to have strong oral communication abilities.

Answers to questions about a candidate's grade point average in school can reveal some possible information concerning a candidate's overall intellectual level. However, to interpret this information properly, an interviewer needs to gather some additional information. This information includes: (1) how much someone studied; (2) how heavy a candidate's course load was; (3) the difficulty level of a candidate's curriculum and the school in general; (4) the extent of a candidate's involvement in extracurricular activities; and (5) the number of hours per week that a candidate worked in some position during the school year [4, 5].

If a candidate received extremely high grades in a very difficult major that was pursued in a school with an outstanding academic reputation, this undoubtedly means that the candidate is rather bright. It also typically means that the candidate worked rather hard in school as well, unless he is so intellectually gifted that he could attain very high grades without a great deal of study.

On the other hand, if a candidate received only mediocre grades in school, the factors described above (that is, (1) to (5)) need to be evaluated to make the proper interpretation. For example, if a candidate received only mediocre grades but indicated that she did not study a great deal and that she worked in a job forty hours per week during the school year during her entire college career, this does not necessarily reflect low overall intellectual ability.

Knowing something about a candidate's extracurricular activities in college can give the interviewer some information concerning the candidate's interest in others [5]. People who are involved in numerous extracurricular activities are often rather gregarious, extroverted individuals. On the other hand, some of the people who are not involved in any extracurricular activities in college may tend to be a bit more reserved and introverted. Candidates who report having been involved in leadership positions in various organizations while in school are often people who have demonstrated an early interest in leadership and took the opportunity to develop their leadership skills to some extent while pursuing their education.

A candidate for a technical position who reports having worked a significant number of hours per week in a job during the school year has some advantages over a candidate who did not work while in school. First, the former candidate had to have developed his time management abilities to some extent to be able to hold down a job while going to school full-time [5]. Such a candidate may also have had the opportunity to develop his work ethic to a greater extent than the candidate who did not have a job while in school [5].

A candidate who has been able to maintain an effective balance regarding studying, participation in extracurricular activities, and part-time work while in college often has the opportunity to develop in an overall sense to a greater extent than does a candidate who focuses the majority of her efforts on one of the three aforementioned areas [4]. As a result, the former candidate may often be more desirable for some technical positions than would the latter individual.

For example, suppose a product design engineer position involves considerable customer contact. A candidate who spent nearly all of his time in college studying might not feel entirely comfortable in developing relationships with customers. On the other hand, another candidate who divided her time in college between studying, participating in extracurricular activities, and part-time work might have developed her interpersonal skills to a greater extent than the first candidate. As a result, she might feel more comfortable and be more skillful in developing relationships with customers than the first candidate would.

3.5 COMPARING CANDIDATES AND MAKING AN OFFER

After all of the candidates for a position have been interviewed and each candidate has been evaluated on each of the key criteria related to the job specifications, supervisory preferences/expectations, and organizational culture, it is very helpful for a technical manager to do some type of write-up that describes how well each candidate matches the key criteria.

Putting the aforementioned information in writing is sometimes very helpful to a technical manager in clarifying his thinking about each candidate. Based on a review of this information, a technical manager can assign a category to each candidate. These categories could be "not recommended," "recommended with some reservations," "recommended," and "strongly recommended." Alternatively, a technical manager might use a rating scale, ranging from one to four, where each of the whole numbers on the scale would correspond to the aforementioned categories.

Whether the categories alone are used or a technical manager decides to use numerical ratings corresponding to the categories, the bottom line is that a technical manager needs to determine which candidate best matches the job specifications, supervisory preferences/expectations, and the organizational culture.

After the best candidate has been selected, a technical manager needs to determine an appropriate salary offer. "Appropriate" does not mean the lowest possible offer that a candidate might accept. Instead, a technical manager must take into account what the candidate has been making in the past as well as what she is worth in the marketplace and to the manager's organization to arrive at the best figure. As indicated in Chapter 1, if a technical manager truly wants to get outstanding employees, he needs to be willing to pay exceptional salaries.

If a technical manager provides a candidate with a "low ball" offer, he is taking a number of risks. First, the candidate may perceive that the organization is "cheap" and is not willing to pay her what she is truly worth. As a result, a candidate is likely not only to turn down the offer, but may well be unreceptive to additional, even higher offers from the organization. Alternatively, a candidate may accept a "low ball" offer because she feels pressured to do so for a number of reasons (for example, she really wants to leave her current job or she is out of work). However, such a candidate would often begin her employment with a "bad taste in her mouth." A candidate who begins her career with an organization in this manner is not necessarily going to demonstrate the same level of effort and commitment that she might if she perceives the initial pay to be very favorable.

As mentioned previously, a manager should try to do some selling of candidates when they are interviewed. Once an offer has been made, additional selling must be done. A technical manager needs to do an effective job of selling the candidate selected regarding the job, himself, and the organization. As mentioned in Chapter 1, it is probably in the best interests of both the candidate and the organization to bring up not only the positives but also the less desirable aspects of the situation. This "realistic job preview" typically provides the candidate with a generally favorable impression. This is true because she does not feel that the technical manager is trying to mislead her by deliberately avoiding any mention of negative information.

If the Number One candidate turns down the job, a technical manager obviously will go to the next best candidate. It is essential that this or any subsequent candidates all be viewed as individuals who are capable of performing very effectively in the position.

3.6 SOME CONCLUDING COMMENTS

Interpreting the information gathered from employment interviews is complicated. However, effective interpretation is an ability and, as is true for other abilities, can be improved through learning and effective practice.

Some technical managers may simply throw up their hands and say, "People are just too unpredictable and too difficult to read. I think I'll just leave interviewing up to the experts." It is true that the guiding principles concerning the understanding of interview information are not as definitive as are various principles of physics, chemistry, and mathematics. Despite this, however, interview interpretation ability can be improved, and it is essential that all technical managers make every effort to do this.

Reading this chapter is a good start in terms of improving a technical manager's interviewing abilities. However, it is only a beginning. To develop interviewing abilities to a significant extent, a technical manager needs to become a student of interviewing. He should go to the library or the local bookstore and borrow or buy a good book on employment interviewing, such as those by Smart and Fear [4, 5].

A technical manager should also take one or more seminars in employment interviewing. Various organizations involved in the development of managers, such as the American Management Association, offer such seminars. In addition, a technical manager can check with a local university and/or technical college concerning the offering of interview courses. I also offer a one-day seminar on selection interviewing. It is a very hands-on, participative, practical seminar that provides participants with the opportunity to develop their interviewing abilities to a significant extent.

Another way for a technical manager to improve interviewing abilities is to talk to and observe an individual within his organization who is adept at employment interviewing. Such an individual might be able to provide some helpful tips to improve his interviewing abilities. In addition, she may be willing to allow a technical manager to observe her doing an actual interview. A follow-up discussion concerning such an interview could be extremely valuable to a technical manager. In addition, a technical manager might ask the interviewer whom he has observed to observe him in one of his interviews and to critique it afterward. Receiving such feedback after conducting an interview can be very instructive and helpful.

Even after a technical manager has done all of the above activities and has had a chance to interview a fair number of actual candidates for open positions, he should never allow himself to feel that he knows everything about employment interviewing. I have a B.A., M.S., and Ph.D. in psychology as well as an M.B.A. in management. In addition, I have had over twenty years of experience in industrial/organizational psychology and human resources, which has included about a thousand in-depth selection interviews of candidates for technical and managerial jobs at all organizational levels. I typically follow up after a person whom I have interviewed has been hired to determine whether my interview observations and interpretations as well as my recommendation concerning hiring the individual have proven to be accurate.

My track record in interviewing has been exceptionally good, and I consider myself an expert in the area. Despite this, however, I continue to learn and improve my interviewing abilities. I am well aware of the fact that I will never know everything there is to know about interviewing, and I can continue to improve my abilities as long as I conduct employment interviews.

Employment interviewing might be considered as much an art as a science. To be skillful at interviewing, an individual needs to have good intuitive abilities, analytical abilities, observational abilities, and overall intellectual capabilities. When appropriate knowledge about interviewing and effective practice are added to the aforementioned abilities, a technical manager can develop into an effective interviewer.

3.7 SUMMARY

In this chapter, I emphasized the importance of effective interview planning and discussed various types of standard and customized questions. Legal issues in interviewing was another topic covered. I then described some key aspects of conducting interviews, including rapport-enhancing techniques and how to conclude an interview effectively. Next, I discussed the interpretation of interview information, including a candidate's work experience and educational background. Finally, I discussed the comparison of candidates and making an offer.

References

[1] Weitzel, James B., *Evaluating Interpersonal Skills in the Job Interview: A Guide for Human Resource Professionals*, New York: Quorum, 1992.

[2] Moffatt, Thomas, L., *Selection Interviewing for Managers*, Madison, WI: Science Tech, 1987.

[3] Goodale, James G., *The Fine Art of Interviewing*, Englewood Cliffs, NJ: Prentice-Hall, 1982.

[4] Smart, Bradford, *Selection Interviewing: A Management Psychologist's Recommended Approach*, New York: Wiley, 1983.

[5] Fear, Richard, *The Evaluation Interview*, Third Edition, New York: McGraw-Hill, 1984.

[6] Orpen, Christopher, "Patterned Behavior Description Interviews vs. Unstructured Interviews: A Comparative Validity Study," *J. Appl. Psych.*, Vol. 70, No. 4, 1985, pp. 774–776.

[7] Kahn, Steven, Barbara Brown, and Brent Zepke, *Personnel Director's Legal Guide*, Boston, MA: Warren, Gorham & Lamont, 1984 (plus 1985 supplement).

[8] Berry, Steven L., et al., *Employment Law: A Complete Reference for Business*, Government Institutes, Rockville, MD, 1993.

[9] Bruce, Stephen D., *Face to Face: Every Manager's Guide to Better Selection Interviewing*, Business & Legal Reports, Madison, CT, 1992.

Chapter 4

The Use of Selection Devices Other Than Interviewing

4.1 INTRODUCTION

Chapter 3 discussed how a technical manager can use the employment interview as a valuable technique for improving the accuracy of selecting employees. This chapter discusses additional techniques that can provide valuable assistance to the technical manager in the selection process. These additional techniques include tests, work samples, reference checks, and assessment centers.

These other techniques can be valuable in the selection process because they can provide important additional information regarding candidates that can supplement what is gained from an interview. Section C.1 of Appendix C compares these additional selection devices regarding a candidate's and a manager's time involvement, whether a consultant is needed, and the approximate cost. More specific details concerning each of the additional selection devices will be provided in the sections that follow.

4.2 TESTS

4.2.1 Types of Tests

The term "tests" is used here to refer to instruments consisting of various items that are administered in a uniform manner. In addition, there is some type of norm group for comparison purposes. For example, the Bennett Mechanical Comprehension Test is a test sold by the Psychological Corporation that provides information on an individual's ability to use mechanical and physical principles in everyday situations [1]. This instrument has uniform directions, and a standard time limit applies to everyone who takes the test. There are also norms provided so that an applicant's score can be compared to the scores of other applicants in various industries.

Strictly speaking, the term "test" applies only to instruments that have right and wrong answers. However, in popular usage, tests also include instruments that do not have specific right or wrong answers, such as personality inventories. My use of the term "test" here is consistent with popular usage.

Within the selection context, tests can be used to measure a number of job-relevant factors. These include knowledge, skills, abilities, and personal attributes. First, a technical manager may be able to employ tests that measure knowledge to improve the accuracy of the selection process [2]. For example, a technical manager may want to ensure that candidates for a particular position have adequate knowledge of the particular industry in which the company operates. Many industries have one or more professional associations that offer examinations (sometimes in conjunction with a course or sometimes separately) that measure one's knowledge of a particular aspect of the industry. A technical manager might ask the candidates for evidence of having passed a specific examination in order to be considered for a particular position. If they have not yet taken the exam but claim to have the knowledge that is measured by the test, a technical manager might request that each candidate take the examination offered by the professional association to ensure that every individual has the knowledge they claim to possess.

A technical manager may also use tests to determine whether the candidates for a particular position possess job-relevant skills. For example, a technical manager may want to hire an individual for an equipment technician position. The job might require that an individual have good skills in using various hand tools. These skills could be measured by certain standardized tests that the manager could use as one element of the selection process.

One point should be added here concerning a technical manager's ability to purchase certain tests. A few tests that are very straightforward in terms of their interpretation require only minimal qualifications for purchase. Other tests, particularly those that are somewhat difficult to interpret, can be purchased only by individuals who hold a license as a psychologist or other advanced qualifications concerning test interpretation [4]. Examples of tests in the latter category include tests that measure job-relevant personal factors (that is, personality inventories) [4].

Candidates' abilities constitute other factors that a technical manager may be able to measure through the use of tests [2]. For example, a product development engineer may want to hire an individual as a lab technician. The engineer might be willing to train such an individual as long as she demonstrates good mechanical ability (assuming this is important regarding the position). As I mentioned previously, commercial tests are available that measure mechanical ability [1]. As a result, the product development manager might want to use such a test in the process of selecting a lab technician.

One last category of tests includes those that measure job-relevant personal factors (that is, personality inventories) [2]. For example, a director of research for a pharmaceutical firm might find that research chemists who work for the organization must persevere in their work, despite experiencing repeated failure. There are certain personality inventories that measure, among other factors, perseverance. As a result, the director of research might ask a doctoral-level psychologist to assist him in the process of selecting research chemists. She could do so by ordering and providing assistance in the interpretation of the results of a personality inventory that includes a measure of perseverance. If, as is probably true for most organizations, there are no internal doctoral-level psychologists who are qualified to interpret personality inventory results, then the director of research might employ the services of a consulting psychologist to assist him in the process of selecting a research chemist.

4.2.2 Why Many Organizations Do Not Use Tests

Many organizations probably do not use tests in the selection process for a number of reasons. First, most technical as well as nontechnical managers do not understand employment tests. Many technical managers seem to believe that either tests are useless or possess some sort of "magical" quality. The truth of the matter is that tests fit neither of these extremes. They are not useless, but neither are they fantastic predictors of job performance. However, tests can be of significant value in the selection process.

A second reason why many technical managers may not use tests is that they are afraid of them. A brief review of some of the historical developments regarding the use of employment tests might shed some light on this fear.

In the 1970s, there began a wave of law suits against organizations for employment discrimination. In some of these cases, employment tests were being challenged. As a result, many organizations simply decided to throw out all testing [3]. Unfortunately, this is an example of "throwing the baby out with the bath water." The ironic part of this is that in some cases, organizations threw out tests that had been properly used (that is, the tests were job-relevant). They also stopped using tests that were capable of improving the accuracy of selecting employees. These potentially superior methods of selection were thrown out in favor of much less effective techniques, such as unstructured interviews, that were not particularly related to the specifications for the open positions.

In recent years, however, there has been a resurgence of testing in employment [2, 3]. Many employers are now probably more sophisticated regarding the use of tests, and as a result, tests are probably being used more appropriately and effectively today than in the past. Despite this resurgence in testing, however, the number of organizations that do not use tests in the selection process is still relatively large. I hope that some technical managers, after reading this chapter, will recognize the value of tests and, as a result, will urge their organizations to begin using testing appropriately to improve their selection process.

4.2.3 How Tests Can Assist a Technical Manager

A technical manager may find that tests can be invaluable regarding measuring certain factors that may be quite difficult to assess via an interview. For example, a vice president of technology for a software company may determine that successful software developers need to have significant creative abilities. Although there are some questions that the vice president of technology might use, a test that measures creativity might be a more accurate indicator of various candidates' creative abilities.

One example of a test that measures creativity is the Remote Associates Test [5]. This test, which is sold by Houghton Mifflin, requires that an individual find a word that is associated with each of three other seemingly unrelated words. People who are very creative tend to get higher scores on this test than do individuals who are considerably less creative. Therefore, candidates for a software developer position could all be given the Remote Associates Test. The various candidates' scores on this test would provide one common index of comparison concerning creativity.

One significant advantage that standardized tests have over alternative methods of assessing certain factors is the ability to make quantitative comparisons of candidates. For example, a director of manufacturing for a medical equipment manufacturer might determine that mechanical ability is a critical variable for equipment repair technicians who are employed by the organization. The director of manufacturing could ensure that all candidates for an equipment repair technician position take a standardized test that measures mechanical ability. Then all the candidates could be compared in terms of their specific score on the test. This would not be the only factor that the director of manufacturing would want to consider in comparing candidates; however, having specific scores concerning mechanical ability would provide one fairly objective means of comparing the various candidates.

4.2.4 Evaluating Various Tests

If a technical manager is interested in determining whether certain tests might be helpful in the selection process and also in comparing various tests that measure a particular job-relevant factor, then hiring an outside expert in testing is probably a good idea. Ideally, the consultant used in this process would not be one who sells tests that are relevant to a technical manager's selection concerns. This is because if a qualified test consultant sells a specific instrument, it might be difficult for her to be totally objective about whether the test that she sells is, in fact, the best instrument to use in a particular situation.

Even though I advise technical managers to employ the services of an independent consultant who specializes in testing, it is a good idea for technical managers to be aware of some of the key issues that should be addressed in the evaluation of tests. One key issue, for example, concerns test reliability. Although I will not go into a detailed explanation here concerning test reliability, a very rudimentary definition of reliability can be provided for technical managers. Reliability simply refers to how consistent a test is in terms of what it measures [6]. Test reliability is typically expressed in terms of a correlation coefficient, *r*, that may range from 0 to 1.0. The closer a test's reliability is to the upper end of this range, the better the test is in terms of its reliability.

Another key measure regarding evaluating tests is the issue of validity. Once again, I do not wish to go into a detailed technical discussion of test validity here; however, it can be simply said that validity involves whether a test measures what it is supposed to measure [2]. In the employment situation, it is important to determine a test's validity for a specific purpose. For example, a test publisher may indicate that a particular test is a valid measure of mechanical ability among high school graduates who have had no advanced technical training. If such individuals represented the only norm group upon which a test's results were based, the test might be very useful in differentiating among candidates for entry-level technical positions that require no advanced training or education. However, if a technical manager wishes to differentiate among graduate engineers in terms of their mechanical ability, such a test might not be particularly valid for this purpose.

Validity is also typically measured in terms of a correlation coefficient. It can range from 0 (no relationship whatsoever) to +1 or –1 (a perfect positive or negative relationship). For example, suppose a test of mechanical ability used with graduate engineers has a validity coefficient of 0.42 and another similar test has a validity coefficient of 0.32. The first test would be a better predictor of mechanical ability than the second.

Another key issue that a technical manager needs to consider regarding testing is the cost [6]. The cost of each test multiplied by the number of individuals who are likely to be tested in a given year provides an estimate of the total annual out-of-pocket costs of using the test.

A related issue that should be addressed concerns the cost:benefit ratio. In other words, if a technical manager uses a consultant to assist with the test evaluation process, the manager should ask the consultant to estimate the savings that are likely to result from the implementation of the test in the selection process. This figure can then be compared with the cost of the test to get an idea of the cost:benefit ratio. For example, if the cost of using a particular test were estimated to be $30,000 a year and the estimated savings in terms of, for example, reduced turnover costs and lower training costs, were $60,000 annually, this would represent a 1:2 benefit ratio. In other words, for every dollar that is spent on testing, $2 are saved.

An additional consideration regarding testing is the time involvement. Some tests might provide similar results and benefits despite being much shorter than other tests.

Another issue a technical manager needs to consider is conducting the tests. The manager needs to determine whether the existing staff can handle the test administration responsibilities or whether additional people would be needed to conduct the tests and compute the scores. I am referring here to tests that are fairly straightforward in terms of administration and scoring. Therefore, the staffing in this case would probably involve clerical/administrative individuals. The cost of any additional staffing would depend upon the number of people needed and the labor rates in the area.

Another important issue that needs to be addressed is what type of training is needed to administer, score, and interpret the tests [6]. As I mentioned earlier, certain more complex tests may require an outside consultant (for example, a consulting psychologist) to interpret them properly.

Finally, a technical manager needs to determine who can purchase the tests. As previously indicated, certain tests can be purchased only by licensed psychologists and others who have had advanced training in the use of tests [4]. Such an individual could purchase certain tests and provide training regarding the administration and scoring of the tests to the appropriate staff members. A qualified consultant could also either interpret the tests or provide appropriate training to a technical manager regarding the proper interpretation. In addition, a consultant should monitor the process to ensure that each test is being used properly within an organization.

4.2.5 Appropriate Use of Tests

Even if a technical manager is using a qualified consultant, it is important for her to be aware of some of the basic issues concerning the appropriate use of tests in organizations.

First, before an organization can make an intelligent decision concerning whether tests are appropriate and, if so, which tests might be best, it is critical to define the various specifications associated with the position, the supervisory preferences/expectations, and the corporate culture (these topics were discussed in Chapter 2). After these three factors have been defined, an assessment can be made concerning whether tests would be helpful in measuring some of these factors. For example, a technical manager may feel that it is important for one of his staff members to handle a position very autonomously, because a manager may employ a supervisory style that is rather "hands off." In other words, the manager may sit down initially with a staff member to define, in general, the objectives regarding a project. After that, the manager might expect the staff member to determine how the project will be accomplished primarily on her own. The manager might have only a limited amount of time to provide a great deal of direction, and/or providing minimal direction may simply be one of his supervisory preferences.

There are certain tests that measure one's ability to operate autonomously, in addition to other factors. For example, the 16PF, which is sold by the Institute for Personality and Ability Testing [4], has 185 items, each of which has three multiple-choice response options. Sixteen primary personality factor scale scores and five global factor scale scores can be generated for each test taker. One of the primary factor scales and one of the global factor scales relate to the ability to function independently. Knowing a candidate's score on these two scales would be helpful in determining whether he would be able to operate autonomously.

An appropriate use of tests in this case would involve evaluating the 16PF versus other personality inventories that might also measure an individual's ability to function autonomously. Once a specific test is selected for use, staff members would need to be trained regarding the proper test administration and scoring procedures. The test results would need to be interpreted by a qualified individual. Also, the administration, scoring, and interpretation procedures would need to be monitored from time to time to ensure that the recommended procedures are being followed.

If the above example highlights the appropriate and effective use of a test, what are some situations that might illustrate inappropriate uses of tests? Using the above example, a technical manager should recognize that if any of the aforementioned elements are missing in the process, this may well be an example of an inappropriate use of a test.

One specific example of an inappropriate use of a test involves one organization that used a test that measures overall intelligence. Undoubtedly, an organization could probably justify the relevance of certain intellectual capacities in almost any position. However, it should be obvious that what is required in terms of overall intelligence in one position is not necessarily relevant in another position. For example, when hiring an engineering technician, it is rather unlikely that the person would need to be as bright as the vice president of technology. Although this last point may appear to be rather obvious, it apparently was not evident to this particular organization, because the company had a single minimum test score that was applied to all employees. This particular use of the test was clearly inappropriate.

4.2.6 Benefits of Using Tests in Selection

Using tests can dramatically improve the accuracy of the selection process [2]. For example, I did a test validation study for one organization that experienced significant turnover in one of its key positions. The results of the study suggested that by using scores on two personality inventory scales, the organization would be able to reduce its poor hires in the key position by about 9 percent.

In addition, an added potential benefit of using tests is a significant reduction in the various costs associated with hiring (for example, the cost of recruiting services or advertisements, the cost of employee training, the opportunity costs associated with having a position unfilled for a period of time, and the cost of reduced productivity due to turnover) [2]. For example, the results of the study previously mentioned suggested that the organization could save approximately $4 million annually by using two personality inventories in the selection process.

In another case, which was reported by Lawshe and Balma [6], an organization that had not been using any tests began to use one in their selection process for a key position. After this single test was added to the selection process, the training costs for the key position were reduced by 39 percent.

The appropriate use of tests in the selection process also may help an employer defend itself against lawsuits regarding discrimination in employment. This is true because evidence of the use of a specific, quantitative measure that is reliable, job-relevant, and valid for the particular position in question can carry considerably more weight in legal proceedings than would the use of nonquantitative, subjective, haphazard, undocumented methods for selecting employees.

Using tests appropriately in the selection process also can help to improve productivity [2]. Since test usage can help an organization to avoid making mistakes in hiring, the net effect is that the overall quality of employees increases. Hiring better employees translates into improved productivity in each position for which tests are used in the selection process. These improvements in productivity in individual positions, in turn, when added up for an entire organization, can easily translate into a significant impact on the bottom line.

To illustrate how tests can be used to improve productivity, Lawshe and Balma [6] reported a study concerning lab technicians in which four tests were used to predict actual job performance. The researchers found that if an individual's scores on the four tests were above certain "critical scores," the probability of the individual being an average or better performer was 90 percent. On the other hand, if a person scored below the critical scores, his chances of being an average or better performer were only 18 percent. Thus, the tests were extremely useful in predicting job performance. As a result, by using the four tests in selection, the organization was in a position to hire better performers and to improve the average productivity of their lab technicians.

4.2.7 Potential Drawbacks of Testing

Paradoxically, although using job-relevant tests in an appropriate fashion can serve as a potential legal defense for an employer who is sued on the basis of employment discrimination, using tests may also tend to increase the likelihood of lawsuits. This probably results from the fact that most people do not understand tests, and some individuals may be afraid of them. As a result, people may sometimes complain about an invasion of privacy when they are asked to take employment tests [2]. In my experience, however, candidates who refuse to take tests as part of the selection process are very rare.

In addition, some individuals may find certain items objectionable, even though the test may be job-relevant and appropriately used regarding selection for a particular position. I find it rather amusing that some people will challenge the appropriateness of a job-relevant test in an employment selection context, yet these same people will not challenge the use of an informal, haphazard, unreliable, invalid interview that might contain a number of questions that are not particularly job-relevant.

Another potential drawback of the use of tests is that certain candidates may feel so strongly about not wanting to take employment tests that they may consequentially remove themselves from further consideration regarding a position. Although a few potentially decent candidates might be lost in this manner, my overall feeling is that very few truly outstanding individuals are likely to remove themselves from consideration because employment tests are used. In fact, a number of candidates for engineering and managerial positions with whom I have spoken in the past have been favorably impressed in situations where a company has taken the time to develop a formalized process involving the use of job-relevant tests to improve the accuracy of selection. Such individuals often feel that an organization that is more sophisticated and selective regarding its process of choosing employees is one that they are happy to have the opportunity to join.

4.3 WORK SAMPLES

Another type of selection device that can be very helpful in improving the effectiveness of an organization's overall selection process is known as a work sample [2, 7]. As the name implies, this approach involves getting a sample of the candidate's work-related behavior. For example, the director of engineering consulting for a management consulting firm might ask candidates for an engineering consulting position to provide samples of past client reports that they had written. In this case, if all of the candidates had been in consulting previously and could provide samples of their previous client reports, this work sample could be very useful as one aspect of the selection process.

One potential drawback of using a work sample that is based on something a client has done in the past, however, is that a technical manager cannot guarantee that the work sample was truly generated by the candidate. For example, in the previous illustration concerning the consulting engineering position, one or more of the candidates might submit a client report that was written in part, or in its entirety, by someone else. To control for this possibility, a technical manager can ask candidates to provide work samples while they are visiting the prospective employer. For example, a director of manufacturing engineering might show candidates a particular manufacturing process that has been causing some problems in the past. He might then ask candidates to spend some time analyzing the process described and demonstrated and then request that candidates develop some written recommendations concerning improving the problematic process. The various written recommendations developed by the candidates can then be compared to one another as well as to the recommendations that had been developed by the organization's existing staff. If the organization actually uses any of the candidate's recommendations, it would probably be appropriate to provide them with some type of compensation.

When using a work sample as a selection device, a technical manager must be sure to request a similar work sample from all candidates. For example, in the previous illustration involving the problematic manufacturing process, it would be inappropriate and unfair to ask some candidates to make recommendations concerning improving that process while asking other candidates to make recommendations concerning the improvement of a different manufacturing process. Section C.2 of Appendix C provides some examples of work samples that might be used with regard to certain selected positions.

4.4 REFERENCE CHECKS

Another device that can aid the overall selection process considerably is the reference check [8–10]. I have heard various individuals malign the value of reference checking. Some of their complaints include: (1) people put down only the names of references who will say positive things about them; (2) you can't get references to say anything negative about candidates; and (3) people aren't willing to serve as references any more because they are too concerned about the possibilities of a lawsuit.

I would like to address each of these three criticisms. First, it is true that most people, when asked to give a list of references, provide the names of only those people whom they expect will give favorable information about them. As a result, I do not recommend that technical managers ask for a list of references. Instead, a technical manager can simply indicate that she would like to talk to the candidate's previous two supervisors, and the manager can then ask for the candidate's permission to do so. If the candidate is currently employed, he usually does not want a potential employer to talk to his current employer. In this case, however, a technical manager can request the names of previous supervisors who may no longer be working for the candidate's current employer. Alternatively, a technical manager can ask the candidate to provide the names of two supervisors with previous employers.

Regarding the second criticism (that is, that people are unwilling to give any negative information about candidates), I have not had much difficulty soliciting unfavorable as well as favorable information when doing reference checks. One way a technical manager can avoid getting only favorable information is to ask about a candidate's strengths as well as her developmental needs [8]. This approach makes most people feel that the reference checker is attempting to get a balanced view of the candidate. After having given four or five strengths that the candidate possesses, most people are willing to provide at least one or two areas that could use some improvement [8].

One suggestion that I have with regard to getting more information concerning developmental needs when doing a reference check is to divide the questions in this part of the reference check into two sections. In the first section, a technical manager can simply ask an open-ended question about a candidate's developmental needs. In the second section, a technical manager can ask about specific developmental needs that he has identified through the interviewing and/or testing processes. While a few individuals may be reluctant to say too much regarding a candidate's developmental needs in response to an open-ended question, most people will admit that a candidate has a developmental need if a reference checker specifically asks about it.

Regarding the third criticism concerning reference checks (that is, that people are unwilling to give out information because of concern about potential lawsuits), I have found this to be the exception rather than the rule. Although I have certainly encountered some situations in which an individual has told me that he would not provide any reference information because it was against the policy of his firm, about 80 percent of the people with whom I talk about reference checks are willing to provide information. Often, a candidate's previous supervisor is willing to give some information about her even though her organization has a policy prohibiting this [8].

One additional potential drawback of reference checking that is not mentioned above concerns the credibility of the source of the reference information [9, 10]. A technical manager may receive some negative information about a candidate and may wonder whether the information is quite accurate or somewhat distorted by the perception of the former supervisor.

For example, a candidate may indicate that his performance was hampered by the fact that his former supervisor was virtually never available to provide assistance or guidance when requested. In addition, a candidate may indicate that when he did make decisions on his own, which was quite often due to the unavailability of his supervisor, his boss seldom supported his decision, particularly if anyone had questioned it.

In this case, a technical manager might contact the candidate's former supervisor and that person might indicate that he did not feel that the candidate was a very good performer because she was not capable of functioning autonomously. In this situation, a technical manager might well wonder whether the candidate's or the former supervisor's perception was distorted. Although the candidate might have an incentive to put his performance in a former situation in a favorable light, the supervisor might also have an incentive to try to put the blame for his supervisory inadequacies upon the former staff member.

Unfortunately, there is no good way to determine whether the truth is what is described by the candidate or the former supervisor unless another disinterested party is available for comment [9]. When I have done reference checks in the past and encountered a situation similar to that just described, and the point of view of a disinterested third party is unavailable, I typically make the assumption that the truth probably lies somewhere between the two different perceptions.

If a technical manager does a reference check and finds a situation that is unclear, such as that described above, the manager might simply try talking to several other former supervisors [10]. If upon being pressed for confirmation of the performance weakness identified by one former supervisor, other previous supervisors report no similar performance deficiency in a candidate, a technical manager might conclude that the performance deficiency mentioned by the one supervisor is somewhat of an isolated situation. If, on the other hand, the other former supervisors confirm the performance deficiency, then a technical manager could conclude that this deficiency might be likely to occur in his situation as well.

In Section C.3 of Appendix C, selected questions for conducting a reference check are indicated. This format, although relatively brief, provides some valuable information about a candidate. In conjunction with the other information obtained through interviews and tests, it can be used to aid a technical manager in determining whether a candidate is a good match with the job specifications, supervisory preferences/expectations, and the organizational/departmental culture.

4.5 ASSESSMENT CENTERS

4.5.1 Description of Assessment Centers

Another selection device that is not used by very many organizations in terms of selecting technical employees, but that is quite relevant, is known as the assessment center [2, 3, 7, 11–16]. If a technical manager finds that she is often hiring multiple people for a particular technical position, an assessment center might be very appropriate and provide some valuable information about the candidates.

An assessment center is a selection device that provides a considerable amount of data that can be used in making a selection decision. Assessment centers usually involve a small number of candidates and a small number of "observers." The observers are typically individuals who are very familiar with the position for which the candidates are being considered. They may be successful individuals within the organization who are in, or were in, the position, or they may be individuals who supervise, or have supervised, individuals in the job.

Assessment centers represent a fairly significant time commitment on the part of the candidates as well as the observers. They may run from one to five days in length [12]. During this time, the observers, who must be trained regarding their roles prior to the session, observe the candidates while they are partaking in the various activities. Often, the observers are asked to focus their observation primarily on several of the participants as well as to note any relevant behavior among the other participants. The activities may include "leaderless" group discussions; simulation exercises (that is, "games"); an in-depth, structured interview focusing on the candidate's work history; and some type of in-depth exercise that requires the candidate to go through what might be a typical "in basket" for someone in the position for which they are being considered [12]. The participants are often asked to prioritize the various issues that are included in the in-basket exercise and to perform various activities to deal with these issues (for example, to set up meetings with various people, to write memos on various subjects, or to analyze certain problem situations).

While assessment centers were initially used in business to select from among candidates for supervisory positions, the concept applies also to various technical positions [13]. Designing an assessment center is a fairly complex process that requires an excellent understanding of various assessment tools and human behavior in general. It is strongly recommended that a technical manager employ the services of a qualified consultant to design an effective assessment center. Many industrial/organizational psychologists have expertise in the design of assessment centers. Some consulting organizations even have staff members who specialize in this particular field.

Although I do not want to describe the process of assessment center design in detail here, I think it would be helpful to interested technical managers to get an overview of how this is done. First, a qualified consultant would, in conjunction with the relevant technical managers and technical employees, conduct a detailed job analysis. This would involve identifying the key responsibilities in a particular position and then identifying the key factors that are relevant to successful performance of these various duties. After this, various exercises that measure the various relevant factors are either purchased from an organization that markets these tools or designed for this specific situation.

The consultant then typically conducts training sessions for assessment center observers and for someone who can serve as an overall facilitator of the assessment center after the consultant completes the assignment. The consultant may initially monitor the operation of the assessment center to ensure that it is being conducted properly. After this, the facilitator of the center, who typically supervises candidates for technical positions, could direct the operation of future assessment centers.

4.5.2 Benefits of Using an Assessment Center

One of the most significant benefits of using an assessment center in the selection process is that this technique can be excellent for predicting future job success [11, 12, 14]. One reason for this high level of accuracy is that the exercises in a well-constructed assessment center are very relevant to a particular position. In addition, a typical assessment center is from one to five days long. There are many opportunities for evaluating someone's behavior during that time. Few other selection techniques provide as much data on a candidate as does an assessment center; this is probably a major reason why assessment centers can be very good at predicting future job success, as compared to other commonly used selection techniques.

When assessment centers are used, most candidates tend to feel that they have an opportunity to see how they are likely to perform in a particular job. In addition, most assessment center candidates feel that this method provides a very fair, objective means of evaluating one's ability to do the job [11, 15]. When assessment centers are used, few candidates complain that people are promoted on the basis of politics or "who they know."

Another potential benefit of an assessment center is that some candidates, after participating in job simulations typically found in an assessment center, decide that they really do not particularly want to be chosen for a particular position. Obviously, this is an advantage not only to them but also to the organization when this realization can occur prior to someone actually assuming a new position. A candidate can be saved from the potential embarrassment associated with deciding to quit a job that does not match well with his interests and abilities. In addition, an organization can save a considerable among of time and substantial training costs when candidates for a particular job decide that the position is not a good match prior to their being placed in the position.

4.5.3 Some Potential Drawbacks of Assessment Centers

Although an assessment center can help to improve an organization's selection accuracy, there are also some potential drawbacks of this selection technique. First, it is a very time-consuming selection method [2, 16]. Considerable time needs to be spent in developing the exercises for a center and in putting together a complete assessment center program. In addition, a great deal of time must be spent training a facilitator for the center as well as the numerous observers who are needed. In addition, the candidates in an assessment center need to spend a considerable amount of their time.

In addition to the time involvement, which is considerable, developing and running an assessment center can be a fairly costly investment [2, 11, 16]. The primary cost is for an outside consultant to help establish an assessment center. In addition, however, the indirect cost of the total time spent in developing and administering an assessment center is considerable. Regarding the development phase, it would probably not be uncommon for an outside consultant who designs an assessment center to spend 5 to 10 days (or perhaps even longer) training the individual(s) who will administer the assessment center. In addition, all of the individuals who will serve as observers need to be trained for their role. This might typically take five days (or even longer).

Regarding the time involved in the administration of the assessment center, it would not be unusual to take about two days for the 12 assessees, four to five days for the six assessors, and five or more days for the one to two administrators each time the assessment center is conducted [16].

The indirect cost of the time involvement of these people can be calculated by multiplying the total number of days required for each person by her daily salary rate. These figures can then be added for all of the people involved. Obviously, the total indirect cost of peoples' time is considerable.

Besides the indirect cost of the various individuals' time, there are also some direct costs that could be fairly substantial. Probably one of the most significant costs, if candidates are brought in from all parts of the country or world, is travel. There are also some direct costs associated with the purchase of the various assessment center exercises from outside consulting firms that specialize in this area.

4.6 SUMMARY

In addition to interviewing, various other devices can be used to select technical employees. These include tests, work samples, reference checks, and assessment centers. Each of these approaches has advantages and disadvantages. However, if used appropriately, each one can help to improve the accuracy of a technical manager's selection process. Section C.1 of Appendix C provides a comparison among the various other selection devices in terms of the candidate's and manager's time involvement, whether a consultant is needed, and the approximate cost.

References

[1] *Tests and Related Products for Human Resource Assessments*, The Psychological Corporation, San Antonio, TX, 1995.

[2] Arthur, Diane, *Workplace Testing: An Employer's Guide to Policies and Practices,* New York: AMACOM, 1994.

[3] Baehr, Melanie, *Predicting Success in Higher-Level Positions: A Guide to the System for Testing and Evaluation of Potential*, New York: Quorum, 1992.

[4] *Behavior Assessments*, Institute for Personality and Ability Testing, Champaign, IL, 1994.

[5] Tryk, H. Edward, "Assessment in the Study of Creativity," in *Advances in Psychological Assessment,* Volume One, Paul McReynolds (ed.), Palo Alto, CA: Science and Behavior Books, 1968.

[6] Lawshe, C. H., and Michael J. Balma, *Principles of Personnel Testing,* Second Edition, New York: McGraw-Hill, 1966.

[7] Saal, Frank E., and Patrick A. Knight, *Industrial/Organizational Psychology: Science and Practice,* Pacific Grove, CA: Brooks/Cole, 1988.

[8] Smart, Bradford, *Selection Interviewing: A Management Psychologist's Recommended Approach,* New York: John Wiley & Sons, 1983.

[9] Half, Robert, *How to Check References When References Are Hard to Check,* Robert Half International Inc., New York, 1986.

[10] Dobson, Paul, "Reference Reports," in *Assessment and Selection in Organizations: Methods and Practice for Recruitment and Appraisal,* Peter Herriot (ed.), West Sussex, England: John Wiley & Sons, 1989, pp. 455–468.

[11] Feltham, Rob T., "Assessment Centres," in *Assessment and Selection in Organizations: Methods and Practice for Recruitment and Appraisal,* Peter Herriot (ed.), West Sussex, England: John Wiley & Sons, 1989, pp. 401–420.

[12] MacKinnon, Donald W., *An Overview of Assessment Centers,* Center for Creative Leadership, Greensboro, NC, Technical Report No.1, May 1975.

[13] Byham, William C., "Application of the Assessment Center Method," in *Applying the Assessment Center Method,* Joseph L. Moses and William C. Byham (eds.), New York: Pergammon, 1977, pp. 89–126.

[14] Holmes, Douglas, "How and Why Assessment Works," in *Applying the Assessment Center Method,* Joseph L. Moses and William C. Byham (eds.), New York: Pergammon, 1977, pp. 127–142.

[15] Dodd, William E., "Attitudes Towards Assessment Center Programs," in *Applying the Assessment Center Method,* Joseph L. Moses and William C. Byham (eds.), New York: Pergammon, 1977, pp. 161–184.

[16] McCormick, Ernest J., and Daniel R. Ilgen, *Industrial and Organizational Psychology,* Eighth Edition, Englewood Cliffs, NJ: Prentice-Hall, 1985.

Chapter 5

Key Selection Criteria for Technical Employees

5.1 INTRODUCTION

This chapter addresses the topic of key selection criteria for technical employees [1, 2]. A case study is presented for illustration. Many of the criteria that will be discussed with regard to this case study are relevant to various technical positions in numerous organizations. However, it is important to keep in mind the fact that a technical manager needs to define the criteria relevant to each position on an individual basis. In other words, he needs to define for each job the knowledge, skills, abilities, and personal attributes that are needed to do a particular job effectively. In addition, a technical manager needs to define his supervisory preferences/expectations, as well as the departmental and/or organizational culture.

After defining the various criteria relevant to selecting a technical employee for the case study, I present a number of questions that could be used to measure these criteria. A technical manager can use many of these questions in the selection process as long as the criteria measured by the questions are relevant to the particular position for which the technical manager is evaluating candidates. Interviews were discussed in detail in Chapter 3.

5.2 BACKGROUND INFORMATION RELEVANT TO THE CASE STUDY

The position with which I will illustrate some key selection criteria for technical employees is that of product design engineer. The primary responsibilities associated with this position include:

1. Designing new products as necessary;
2. Revise existing products as needed;
3. Supervising technical assistants involved in the product design process;
4. Coordinating product design and revision with the manufacturing department;
5. Assisting the marketing department by making presentations to potential new customers;
6. Communicating orally and in writing with existing customers regarding the product design process.

The technical manager in this fictitious case study is a very demanding individual who sets high expectations for his staff members. He is a very capable individual who expects his staff members to be competent as well. Although he wants his staff members to be able to function with a fair amount of autonomy, he also expects them to be willing to accept direction when necessary.

The organization in this fictional case study is a publicly owned company that generates about $25 million in annual sales. The company's primary products are pressure sensors, and it sells its products mainly to the automotive industry. The company's major customers are very demanding. They have high expectations concerning product quality and the time involved in the product design process.

5.3 KEY SELECTION CRITERIA FOR THE PRODUCT DESIGN ENGINEER POSITION

The primary selection criteria for the product design engineer position include knowledge, skills, abilities, and personal attributes. The criteria also reflect major personal preferences/expectations of the technical manager as well as key elements of the departmental/organizational culture found in this company. The various criteria are not listed in order of importance or in any other particular order.

The key selection criteria for the product design engineer position in this fictional organization include the following:

1. *Knowledge of product design in general.* A top candidate must understand the essential elements of the product design process.
2. *Knowledge of the automotive industry.* An excellent candidate needs to understand the key elements of the automotive industry.
3. *Ability to design pressure sensors for automotive applications.* A top candidate must be capable of designing state-of-the-art pressure sensors for cars and trucks.
4. *Ability to interface with the manufacturing area.* An outstanding candidate must be capable of working closely with individuals from the manufac-

turing function in the product design process. Individuals need to be able to design products "on the cutting edge," but they must also ensure that products can be manufactured easily and at a reasonable cost.

5. *Ability to manage multiple projects simultaneously.* An excellent candidate must be capable of managing the product design process for multiple new products simultaneously. In addition, the individual must be capable of managing the design revision process for multiple existing products at the same time.
6. *Ability to supervise technical assistants.* An outstanding candidate needs to be able to manage her staff members effectively.
7. *Work ethic.* An excellent candidate for the position must be willing to work long hours on a regular basis to complete all of the job tasks.
8. *Sense of urgency.* An outstanding candidate must want to get things done as soon as possible. The individual must be able to meet challenging deadlines on a regular basis.
9. *Perseverance.* Individuals who are outstanding candidates for this position cannot give up easily. They must be capable of being persistent, even in the face of adversity. However, they must also be capable of recognizing when a particular course of action is ineffective and be willing to change direction in such cases.
10. *Conscientiousness.* Good candidates must be very responsible, dependable individuals who are willing to do whatever is necessary to meet their responsibilities.
11. *Self-confidence.* Viable candidates for this position need to be confident of their abilities in general, and they must be capable of conveying this confidence to others. However, they cannot come across in a condescending or arrogant fashion.
12. *Assertiveness.* Top candidates need to be capable of "pushing" when necessary to get the job done. However, they also need to know when to "back off," and they cannot come across as being abrasive.
13. *Intelligence.* Excellent candidates need to be capable of solving complex technical problems. In addition, they need to have considerable common sense and be able to solve everyday practical problems as well.
14. *Creativity.* Outstanding candidates for this position need to be capable of generating novel product design ideas.
15. *Ability to communicate.* Top candidates must be able to communicate effectively orally and in writing with customers as well as other employees. The key here is to be thorough, but also concise in communication. The individual should avoid jargon as much as possible.
16. *Ability to function relatively autonomously.* A top candidate should be capable of functioning on her own; the individual should need minimal guidance from the product design manager, with the exception of direction concerning overall objectives. However, she cannot demand total

autonomy; an excellent candidate must be willing to accept some direction from the product design manager.

5.4 SAMPLE INTERVIEW QUESTIONS FOR THE PRODUCT DESIGN ENGINEER POSITION

In this section, I will present a number of interview questions that might be used to select a product design engineer from among candidates for the position. The questions are not listed in order of importance. Many of these questions were taken from the list of sample standard questions presented in Chapter 3. In addition, I developed a number of other questions to supplement the standard questions. I recommend this same approach to technical managers who are planning to interview job candidates.

The questions presented in Appendix D are designed for the specific product design engineer position in this case study. However, as I mentioned previously, many of the criteria that are indicated for this sample position might also be relevant to various positions for which a technical manager might be interviewing candidates. As a result, many of these questions would also be relevant in cases in which the criteria upon which the questions are based are job-relevant.

One selection criterion for the position is knowledge of the product design process in general. Question 1(a) of Appendix D relates to this criterion. It asks a candidate to rate himself regarding each of the technical aspects of the position. Question 1(b) asks him to provide evidence supporting each rating. If a candidate does not mention product design in response to question 1(a), a technical manager can ask a specific question on this topic. Candidates who rate themselves highly in terms of their knowledge of the product design process and are able to justify the high rating with meaningful supporting evidence are probably quite knowledgeable concerning product design.

Knowledge of the auto industry is also very important in this job. Information on this criterion can be gathered by asking the same questions asked for the previous criterion. Once again, if a candidate does not mention his knowledge of the auto industry in response to question 1(a), a specific question dealing with this topic can be asked.

The ability to design pressure sensors for automobiles and trucks is critical in this role. Questions 2(a,b) in Appendix D, which relate to a candidate's accomplishments in his positions, are likely to provide information concerning this criterion if a candidate has had previous experience in designing pressure sensors for the automotive industry. If a candidate has designed pressure sensors, but not for the auto industry, this information is also likely to be revealed in response to the questions on accomplishments. Likewise, if a candidate has had no previous experience in designing pressure sensors, this may be indicated by his response to the questions concerning accomplishments.

Obviously, if a candidate has had previous successful experience in designing pressure sensors for the automotive industry, this indicates that a candidate has the ability to do so in this position. What about candidates who have not designed pressure sensors for the automotive industry or those who have not designed pressure sensors at all in the past? For such individuals, a technical manager might simply ask them question 3 in Appendix D. It asks a candidate to provide evidence that he would be able to design pressure sensors for the automotive industry. For example, a candidate might indicate that he has designed other products for the automotive industry and is familiar with pressure sensor technology. In such a case, the individual may well have the ability to design pressure sensors for the automotive industry, even though he has not done so previously.

Another criterion for the product design engineer position is the ability to work closely with the manufacturing function in the product design process. Question 4(a) relates to this criterion. It asks a candidate to provide an example that illustrates his ability to work closely with manufacturing. If a candidate provides a convincing example, this provides one piece of evidence that he can work well with the production area. On the other hand, if a candidate has difficulty coming up with an example, this may suggest that he has not worked closely with manufacturing in the past. This does not necessarily mean that he is not willing to do so, however. A follow-up question, question 4(b) in Appendix D, might provide some information about his willingness to work hand-in-hand with employees from the production area.

The ability to manage multiple projects simultaneously is also important in this job. Question 5 addresses this criterion. It asks a candidate to provide an example that illustrates his ability to manage various projects simultaneously. A candidate might also spontaneously provide some information relevant to this criterion in discussing his accomplishments in the various positions that he has held (questions 2(a,b)).

The ability to supervise technical assistants is another relevant selection criterion. Several questions in Appendix D relate to this criterion. They include questions on a candidate's management style, his rating of himself as a manager, and evidence that supports this rating (questions 6(a–c)). A candidate's response to the question on management style may provide some evidence of his ability to vary his style according to the needs of each particular situation. If a candidate rates himself fairly high as a manager and he is able to provide convincing evidence to support that rating, this suggests that he may have good managerial skills.

The willingness to work long hours to accomplish job-related objectives is also important. Several questions in Appendix D relate to this criterion. Question 7(a) asks the candidate about the time when he starts and ends work on a typical day. Questions 7(b,c) ask a candidate to indicate the number of hours that he typically works in the evening and on weekends.

Obviously, a technical manager could simply ask a candidate to indicate the number of hours that he typically works in a week. However, I do not ordinarily ask this question because I have found that candidates typically provide an overinflated estimate in response to this question, either intentionally or unintentionally. As a result, I have found that I can get a much more accurate estimate of the total number of hours ordinarily worked by an individual by asking separate questions about the typical hours worked during the day, in the evening, and on weekends.

Another question that may provide some information concerning this criterion is question 7(d), which asks about the number of hours per week typically worked in part-time jobs during the school year while a candidate was pursuing his education. Obviously, some candidates may not have had to work at all while their classes were in session. In such cases, a candidate's response to the question does not provide any useful information for evaluating this criterion. However, if a candidate did have part-time jobs while in school, the information provided in response to this question can be very useful. For example, if a candidate worked 30 to 40 hours per week during the school year and also spent a great deal of time studying and attaining a high grade point average, this provides evidence that the individual has a strong work ethic.

Another criterion for the position relates to a candidate's sense of urgency and ability to meet difficult deadlines. The questions indicated in Appendix D that deal with a candidate's accomplishments in his various positions (questions 2(a,b)) may reveal some meaningful information regarding this selection criterion. For example, a candidate may indicate that he was able to complete the entire design process for a particular product in three months, as compared to a standard time of six months within the industry. Such a response seems to provide one piece of evidence that the candidate has a strong sense of urgency and is able to meet difficult deadlines.

Perseverance is an additional important criterion relevant to this job. Question 8, which asks a candidate to provide an example of a situation that demonstrates his persistence, can provide some relevant information concerning this criterion. Information that a candidate provides in response to questions concerning job-related accomplishments (questions 2(a,b)) can also be very helpful in trying to assess the candidate's persistence. For example, a candidate might indicate that he continued his efforts to design a particular product, despite experiencing repeated failure in attempting to do so.

Conscientiousness is also relevant to the position. In response to question 9, which addresses this criterion, a candidate might indicate, for example, that he worked 16 days in a row at his job to complete an important project on time. Such a response provides evidence of considerable conscientiousness.

Another key criterion for the product design engineer position is self-confidence. Questions 10(a,b) ask the candidate to rate his self-confidence based on a scale from 1 to 10 and to support this self-rating. Interpreting the candidate's response, obviously higher ratings on self-confidence would be desirable. However, the evidence provided to support a candidate's rating will either strengthen or weaken the candidate's claim.

Questions 16(a,b), which ask for an overall self-appraisal concerning the position, might also elicit responses providing some information regarding the candidate's self-confidence. For example, a candidate might mention self-confidence as one of his strengths. Alternatively, a candidate might also mention self-confidence as an area that needs further development.

An additional key criterion is assertiveness. Answers to questions 11(a,b), which ask the candidate to rate his assertiveness and to provide supporting evidence, will provide some information concerning this variable. Questions 16(a,b), which involve a candidate's self-appraisal, might also provide some information concerning a candidate's assertiveness. For example, a candidate might list assertiveness as a strength or as an area that needs further development. These self-appraisal questions are excellent because a candidate's response could potentially provide some information on almost any of the criteria for the position.

Overall intelligence is also important with regard to the position. Questions 12(a,b), which relate to a candidate's grade point average in college/technical college or graduate school (if relevant), can provide some information concerning this criterion. However, as noted in Chapter 3, a technical manager usually needs additional information besides a candidate's grade point average to make an accurate assessment of the individual's overall intelligence. These other factors include how much someone studied, the number of hours of work per week that a candidate put in with regard to a part-time job (if he had one during the school year), the extent of the candidate's involvement in extra-curricular activities, the difficulty of the individual's major in school, the number of credits taken each semester, and the overall academic difficulty of the school(s) attended [2, 3]. For example, suppose a candidate indicates that he had only a 2.0 grade point average at a rather mediocre engineering school. If the candidate studied seven hours per day, this individual may not be as bright as some of the other candidates who might be interviewed for the position.

Another question relating to a candidate's overall intelligence is indicated in Appendix D, which asks a candidate to compare himself/herself to others in terms of intellectual ability (Question 12c). If a candidate says that he/she is extremely bright in comparison to others, it is usually a good idea to ask the individual for specific evidence which supports this belief.

An additional key criterion regarding the product design engineer position is creativity. Question 13 attempts to assess this criterion by asking a candidate to provide evidence of his creativity. Questions 2(a,b), which concern a candidate's accomplishments in each of his positions, also might reveal some information with regard to the candidate's creativity. For example, if a candidate indicated that he designed 15 new products in the past five years, this would appear to be evidence that the candidate is rather creative. The questions relating to accomplishments are very good because they may reveal information about a number of the other criteria related to the job.

The ability to communicate well (both orally and in writing) is also relevant in this job. Questions 14(a–d) ask a candidate to rate his oral and written communication ability and to provide supporting evidence for the ratings.

A technical manager can also assess a candidate's oral communication ability by noting how well the individual presents his ideas in the interview. An additional means of evaluating a candidate's written communication ability involves asking the candidate to bring a sample of his writing. This might provide a good measure of someone's written communication ability. However, one disadvantage of this approach is that a technical manager cannot be certain that the writing sample was actually written by the candidate.

Another selection criterion for the product design engineer position is the ability to function fairly autonomously, coupled with a willingness to take direction when necessary. Question 15(a) can be used to assess this criterion by asking about a candidate's expectations from his supervisor. If a candidate needs to be overly dependent upon his supervisor, this may be revealed in the individual's answer to the question. On the other hand, if a candidate is so fiercely independent that he is unwilling to listen to anyone, this may also be revealed in his response to the question.

Question 15(b) also relates to this criterion. It asks a candidate to provide an example illustrating his ability to operate autonomously. An additional question that relates to this criterion (question 15(c)) asks a candidate to provide an example that illustrates his ability to take direction when necessary. The ability of a candidate to provide examples in response to these two questions, as well as the quality of these examples, may provide some information regarding this criterion for selection.

5.5 SUMMARY

In this chapter, I described a product design engineer position in a fictional organization to illustrate various key selection criteria for technical positions. Although each technical position needs to be analyzed individually to determine the major selection criteria, many of the criteria indicated for the job in this case study are relevant to numerous technical roles.

In the associated Appendix D, I provided a list of questions that a technical manager might use when interviewing candidates for the product design engineer position. Once again, each position needs to be evaluated individually to determine appropriate job-related questions that can be used in an interview. However, many of the questions indicated in this chapter and the appendix would be relevant to numerous other technical positions.

References

[1] Coss, Frank, *Recruitment Advertising,* American Management Association, New York, 1968.

[2] Fear, Richard, *The Evaluation Interview,* Third Edition, New York: McGraw-Hill, 1984.

[3] Smart, Bradford D., *Selection Interviewing: A Management Psychologist's Recommended Approach,* New York: John Wiley & Sons, 1983.

Chapter 6

One-on-One Development

6.1 CAN PEOPLE CHANGE?

A technical manager from one of my client organizations asked me a very good question a while ago. Via a program that I had introduced in his company, he had gotten feedback from his coworkers about some changes in his approach that they felt would be beneficial. I reinforced the message from his fellow employees, and he said, "I can see how making these changes would be helpful, but I have been operating this way for over 40 years. Can a person really change his personality?"

I told him that changing his personality might require lengthy psychotherapy and a great deal of motivation on his part, and even then the chances of success were certainly not guaranteed. I further explained that neither I, nor his coworkers, really wanted him to change his personality. Rather, we wanted to see certain changes in his behavior in some specific situations. These changes would help him to improve his performance, and they would also help to increase the overall effectiveness of the work group. Unlike changing his personality, these changes could realistically be made almost immediately, if he chose to do so.

Although changing one's work-related behavior is significantly less difficult than changing one's personality, it still is not easy. Assuming that a person is capable of making a given change, it will occur only if the individual believes that the benefits of the change outweigh the costs (for example, the feeling of discomfort associated with trying something that she has not done before).

Developing an employee implies changing him in some way, presumably for the better. A key point for a manager of technical employees to recognize is that change or development is largely up to the employee. However, what a manager can do is create an environment that encourages an employee to want to change or develop and to support him in this effort.

6.2 KEY ROLE OF THE MANAGER OF TECHNICAL EMPLOYEES

In most of the organizations with which I have worked, many managers of engineers and other technical employees seem to view development as something that "someone else" does. In other words, they feel that they need to send someone to an internal training course or to an outside course or seminar to develop her. Although these options may be quite appropriate in many situations, a manager needs to realize that he plays a critical role in employee development [1, 2]. A manager's role is probably secondary only to that of an employee herself in most situations.

6.2.1 Defining Key Developmental Needs

First, a manager of technical employees typically plays a key part in determining employees' major developmental needs. One of the most important functions of a traditional performance appraisal is to identify areas needing improvement [3]. When writing an appraisal, it is critical that a manager spend a significant amount of time on spelling out an employee's key developmental needs in as much detail as possible.

For example, indicating that an entry-level R&D engineer needs to improve her self-confidence is extremely vague and not particularly helpful. An example of a much more specific developmental need statement is the following, which I suggested to an individual who worked for the Parker Pen Company a number of years ago: "To indicate 'I can do this' instead of 'I don't think I can do this' when given an assignment that I have not previously done."

I also strongly recommend that a self-appraisal be included as a part of the overall appraisal process. This gives a technical employee a chance to identify developmental needs from her own perspective. When these are added to those generated by her supervisor, a fairly comprehensive list of areas needing development should result.

6.2.2 Locating the Right Developmental Vehicle

A technical employee should bear the main responsibility for finding the proper vehicle for meeting a specific developmental need. However, the individual's supervisor should also play a key role. The main reason is simple: "Two heads are better than one." If both the employee and her supervisor are looking, the chances of finding the right means of meeting a developmental need are greatly increased. For example, when I was vice president of employee development for Sentry Insurance, one of my staff members assumed responsibility for succession planning. However, he had not had any formal training in this area, so we both began looking for an appropriate seminar. I managed to find one offered by the Conference Board that looked appropriate for myself as well, so we both attended.

Another reason why a manager should be involved in locating the right developmental vehicle is that many times he may know of some developmental opportunities that his staff member does not. For example, a former managerial-level staff member of mine at Sentry Insurance belonged to the *Insurance Company Education Directors Society* (ICEDS). He was aware of a presentation on sales training that was quite relevant to the developmental needs of one of his staff members. This individual was able to sit in on the presentation and learn from it. Since the staff member was not a member of ICEDS, he would not have been able to take advantage of this opportunity had his supervisor not been looking for developmental vehicles.

Defining a specific developmental need and then looking for a means to meet it is substantially more effective than the approach used in many organizations. Often, a technical employee has had no specific discussions with his supervisor about any particular developmental needs. Nonetheless, upon seeing a brochure on a seminar that looks interesting, the employee asks his supervisor about the possibility of attending.

For example, when I worked for the Parker Pen Company, a manager in the R&D area asked me if I knew anything about a particular seminar on finance for managers not involved in the financial function. I told him that I was familiar with the organization sponsoring the seminar and that it had a good reputation, but that was not the critical issue. The key question concerned whether gaining financial knowledge was one of his primary developmental needs, as defined jointly by him and his supervisor. He told me that the subject had not been discussed; he just thought that the seminar looked interesting.

There are several problems with the approach used by the individual mentioned above. First, it can be rather expensive to send employees to whatever courses or seminars they find interesting. Most organizations today simply cannot afford to do this. Second, even if an organization puts a limit on outside development (by indicating, for example, that each employee may attend no more than one outside course or seminar per year), the above approach is still not cost-effective because people are not attending outside programs relevant to their *most critical* developmental needs (unless they just accidentally come across first-rate seminars that happen to address these needs). Therefore, an individual should ensure that she focuses on evaluating and attending programs relevant to her most significant developmental needs.

6.2.3 Defining Desired Benefits

Once a technical employee's key developmental needs have been defined and appropriate vehicles for meeting these have been identified, the specific benefits desired need to be determined. Once again, the employee should bear the primary responsibility for doing this. However, the employee's supervisor should provide assistance.

For example, one of my former staff members at Sentry Insurance was selected for a management training specialist role. Although he had some supervisory experience, he had no formal training in management. As a result, his selection was contingent upon his enrollment in the four-part series in management sponsored by the American Institute of Property & Liability Insurers.

The employee and I jointly determined the major benefits he expected to gain from the series before beginning the first course. In this case, they were: (1) to use the knowledge gained to assist him in identifying key management development course topics for Sentry, (2) to use what he learned from the program in the design of Sentry courses, and (3) to teach the four-part series to Sentry employees after he had finished the sequence.

6.2.4 Defining Desired Changes

After a technical employee has participated in a specific course or some other developmental experience, the person and her manager should determine specific behavioral outcomes that are desired. In other words, after taking a certain course, for example, what specific things should the person know, be able to do, and actually implement?

For example, two engineers at one of my former employers attended a program on quality at the Juran Institute. Upon return from the program, they were expected to design a customized "total quality program" for their organization based on the Juran principles.

A second example involves another former coworker of mine. He attended a program on work measurement. After returning from the seminar, he was asked to analyze several key production positions and make specific recommendations regarding the production standards for these jobs.

Some people may find it difficult to identify specific actions they can take after completing a course or other developmental experience. It is true that this often is not easy to do. However, if an individual is unable to come up with any specific ways that he can use what was learned on the job, something is wrong. The problem may center around the course (for example, it did not deliver what it was supposed to). Alternatively, the difficulty may lie with the individual (for example, he did not participate fully enough in the course to gain something usable on the job).

Some individuals may take the attitude that "learning for the sake of learning" is enough. They may feel that we do not need to know exactly how or even whether we will use something we have learned. One R&D engineer with whom I worked at the Parker Pen Company, for example, was interested in pursuing a law degree. When I asked how such a degree would help him in his current position, he was not able to tell me. In addition, he had no particular aspirations for any specific positions in the legal department (for example, patent attorney). He simply felt that gaining knowledge, whether it had any practical application or not, was sufficient to justify the company paying for him to pursue a law degree.

My view, however, is quite a bit more pragmatic. I feel that if a company pays for someone to attend a course, that person owes it to his employer to determine specifically how the material can be applied on the job for the benefit of the company.

6.2.5 Feedback on Progress

After a technical employee has implemented the appropriate actions subsequent to completing a developmental activity, she should update the supervisor on progress that she has made. The person's supervisor should schedule progress report sessions at regular intervals to ensure that he is kept up to date.

A person's supervisor should provide an employee with feedback concerning his observations on progress that has been made (or lack of progress). It is critical for an individual's supervisor to provide positive reinforcement about observed improvements in performance. This positive feedback is likely to encourage an employee to continue her efforts to improve.

For example, suppose a manufacturing engineering manager recently attended a seminar on negotiation skills. He may have indicated that he would use the knowledge gained to negotiate lower wage rates than those initially proposed by temporary technical help firms who supply contract designers when they are needed. Suppose also that his supervisor notices that he negotiates a wage rate for three contract designers with a temporary technical help supplier that is 10 percent less than the rate he negotiated the last time he dealt with this agency regarding employees having similar experience. The supervisor should then compliment the manufacturing engineering manager on his success in negotiation.

Conversely, it is important for a manager to provide constructive criticism for lack of progress. The two previously mentioned people can be used in a second example. Suppose the manufacturing engineering manager hires four contract designers at a rate that is 10 percent more than the rate paid previously for people with similar experience. Suppose also that his supervisor believes that, despite a small general increase in wage rates since the last time, he should have been able to negotiate a wage rate that is no higher or perhaps even somewhat lower than last time. If this is the case, his supervisor should talk to him about it. The supervisor might discuss some suggestions of how what was learned in the seminar may have been used to change the outcome of the wage rate negotiations.

6.2.6 Time Consuming, But Well Worth It

Some managers of technical employees who read the above recommendations about their role in a subordinate's development may say, "I can't afford to spend that much time." My response is, "You can't afford *not* to spend the time."

I have observed and participated in employee development in quite a number of organizations. I feel strongly that one of the most significant reasons why development is often less successful than desired is the lack of involvement by supervisors of participants in developmental programs. When the amount of time and money spent by technical employees who participate in developmental activities is considered, it is quite clear that whatever time is spent by their managers to increase the likelihood of success is well worth it.

6.3 TECHNICAL EMPLOYEE MANAGER AS COUNSELOR

If we were to put a name on the various activities recommended for managers of technical employees in the previous section (for example, helping to identify key developmental needs and assisting in locating appropriate developmental vehicles), "counselor" would probably be as good as any. A counselor is someone who guides and assists another. He can serve as a "sounding board" by allowing another person to bounce problems and ideas off him.

A counselor sometimes gives advice to another, but she does not try to solve all of the problems or provide all of the answers for someone else. Why not? One important reason is that a counselor typically tries to develop the counselee's independence. If, for example, a product development manager solves every problem that one of her product design engineers brings to her, what is the design engineer likely to do in the future? He will probably continue to come to his supervisor for solutions to problems. If, on the other hand, the product development manager expects the product design engineer to solve problems himself, eventually the individual will probably become self-sufficient. This is even more likely to occur when the culture of the department and/or the organization reinforces the importance of self-sufficiency.

Many individuals feel more comfortable in a directive problem-solving role than in a counseling role. The former is usually faster and may give a manager of technical employees a feeling of satisfaction to know that she alone solved a problem. Over the long run, however, a manager will end up spending more time by being directive because her staff members will never develop the independence needed to solve their own problems.

Some people are better counselors than others naturally, but counseling is an ability that can be developed through training [4, 5]. Such training includes developing abilities in listening and learning to respond in ways other than just telling someone how to solve a problem. For example, managers can be taught to "reflect back" the content or feeling conveyed by their technical employees concerning perceived problems [4]. This can help the employees to see the problems from a different, more appropriate perspective.

Various books (for example, [4, 5]) provide information on how to train managers to develop their counseling abilities. Seminars on this topic are also offered by various organizations (for example, the University of Wisconsin-Madison Management Institute).

6.4 TECHNICAL EMPLOYEE MANAGER AS COACH

Another major role that a manager of technical employees should play in the developmental process is that of a "coach." A coach is someone who helps another person develop an ability by explaining how to do it, demonstrating the ability, asking the person to demonstrate and practice the ability, and providing constructive feedback on the individual's performance.

Developing an ability by observing others is known as "imitation learning" or "behavior modeling" [6, 7]. Imitation learning is one of the most common and effective ways that we learn. We all use it often early in life to learn how to walk and talk and continue to use it frequently in school and in our career.

However, most of the imitation learning that we do in our jobs is informal and unstructured. We may note what a very ineffective manager does and try to model the opposite behavior. Alternatively, we may observe what a very competent manager does and try to imitate it. When we try to do this informally, however, we are missing some of the key elements of the formal coaching process (for example, we do not usually get an explanation of the behavior and we do not receive constructive feedback on our progress).

Since imitation learning is such a powerful developmental tool, why not use it in a structured, formal way? A competent manager of technical employees has many abilities that her staff members could learn through formalized imitation learning. For example, when I worked at Sentry Insurance, I used a coworker feedback-based program to develop individual managers. To be able to expand the program to more managers than I could work with myself, I taught several of my staff members how to conduct the program. How did I teach them? I did it through imitation learning. I first explained the program and then asked them to observe me conducting the program several times. Then we reversed roles; they conducted the program and I observed them. After receiving constructive feedback on their performance several times, they were ready to conduct the program competently on their own.

What job-related abilities can people develop through imitation learning? The answer is "almost any." These include technical abilities (for example, how to design a fiber optic sensor and how to properly maintain robotics production equipment), managerial abilities (for example, how to restructure a department and how to monitor others' performance), and interpersonal abilities (for example, how to gain consensus in a group and how to sell your ideas).

Unfortunately, although the use of formalized imitation learning by managerial coaches is an extremely effective developmental tool, it is a greatly underutilized technique. Of the numerous organizations with which I have worked over the years, I can recall very few instances where formalized behavior modeling was used. Why is it not used more? I believe that one reason is that organizations do not recognize just how powerful this relatively untapped resource can be. Another reason may be that some managers of technical employees do not know how to coach or do not feel comfortable in this role. In addition, coaching may not be perceived as a high priority by many managers, especially as compared to other critical issues and problems. Perhaps the most significant reason why more coaching does not occur in organizations is that is time consuming.

How then, can this tool become a more significant developmental resource in organizations? First, all levels of managers of technical employees need to be educated and convinced of the value of formalized coaching. Second, coaching skills need to be taught formally in organizations (using internal professional trainers or qualified outside resources). Finally, managers of technical employees need to be persuaded that they cannot afford *not* to take the time to coach their staff members.

6.5 ABILITY DEVELOPMENT TEAMS

6.5.1 Introduction

Although coaching can be an invaluable developmental technique, what happens when an ability that an employee needs to develop is one that the person's supervisor is not particularly adept at demonstrating? Obviously, the manager can use a source outside the organization. This may work fairly well in certain instances (for example, when the manager is already familiar with a highly competent outside resource person or organization and the cost is quite reasonable). However, if a manager is not aware of any well-qualified outside resources in a particular area, he may need to spend quite a bit of time trying to find what is needed. If a manager has to do this concerning several key abilities for each of his staff members, the time required would probably be prohibitive. In addition, the cost of using outside resources for this purpose for all technical employees would probably be unacceptably high for most companies.

A more appropriate solution to the problem may be for a manager of technical employees to enlist the services of what I call an "inhouse ability development team."

6.5.2 How an Ability Development Team Might Work

The following example illustrates how an *ability development team* (ADT) might operate. Let us assume that Judi is the director of product development for Acme Telecommunications, Inc. (a fictional company). Bill, a senior product design engineer, works for Judi. Technically, Bill is the best engineer in the department. However, he needs to develop some of his personal and interpersonal abilities.

Judi, in conjunction with Bill, has defined four key areas in which Bill needs to develop. These areas are as follows: resolving conflict directly, selling his ideas, conducting meetings, and managing time effectively. Judi is widely recognized throughout her company as being highly competent in conducting meetings. However, her abilities in the other three areas range from below average to average.

To help Bill develop in the other three areas, Judi could check the company's *Catalog of Ability Development.* This document lists various abilities that can be enhanced by internal "ability developers" who are highly proficient in each of the areas. In this case, Judi himself, as well as several other people, might be listed under "conducting meetings." Tom might be one of the people indicated under the "resolving conflict" category. Harry could be one of several individuals skilled in "selling ideas." Finally, Michele might be one of four people listed who are proficient in "time management."

Judi could contact Tom, Harry, and Michele and request that they, in conjunction with her, serve as Bill's "ability development team." In this case, Judi might play two roles. As an ability developer herself, she could help Bill to develop his abilities in conducting meetings. As Bills supervisor, she could coordinate the efforts of the other three people in developing Bill's abilities in the other areas. This coordination might include, for example, introducing Bill to each of the others, establishing a schedule for Bill to meet with each of the others, and following up on Bill's progress.

When Bill has developed a particular ability to the desired extent (as defined by himself, Judi, and the ability developer), that ability developer would drop out of the ADT. When Bill has developed the requisite expertise in all four of the abilities, the ADT would be disbanded.

6.5.3 Catalog of Ability Developers

How could the *Catalog of Ability Developers* be produced? A qualified member of the human resources group might be the project coordinator. This individual could first review all of the company's performance appraisal documents and then generate a list of all of the abilities needing development. This document could then be routed to the management group, and individuals could recommend ability developers who are highly proficient in one or more of the areas indicated.

If there is no one in the human resources group who is qualified and willing to serve as a project coordinator or if there is no interest to create a companywide catalog, a qualified technical employee could be selected. This individual could develop a catalog for technical employees in a fashion similar to that described above.

Serving as an ability developer for one or more individuals would obviously take some time. However, I believe that most nominated individuals would probably agree to serve; it would be quite an honor to be designated as one of a select group of individuals who are all experts in a particular ability area. The company could provide some sort of special recognition to ability developers (for example, an annual recognition dinner), but this may not be necessary. Just the honor of being chosen as an expert and experiencing the gratification of developing another person might well be sufficient motivation for the majority of ability developers.

The company could establish some guidelines on the maximum number of areas of expertise (for example, two) and people on which an individual may work at one time (for example, two), so that serving as an ability developer would not become an onerous task. It should be voluntary rather than mandatory to serve as an ability developer because someone forced into service would probably typically not perform as well as would a person who volunteers.

Undoubtedly there would be some ability development areas identified in the review of the performance appraisals in which no internal expert could be found. These abilities might be best developed using outside experts. Likewise, if there is a mismatch concerning the number of people needing development in a particular area and the available internal resources, a company could go to the outside.

ADTs should not be expected to handle all of the development not done by supervisors. However, they can obviate the need for a company to spend a great deal of money on outside developmental resources. ADTs also have another major advantage over many outside resources; that is, they can tailor ability development to fit the unique requirements and culture of a given company. For example, "going over someone's head" (that is, going to one's supervisor) to resolve a dispute may be quite common and considered acceptable in one organization, but it might be considered totally unacceptable in another.

6.6 USING COWORKER FEEDBACK IN INDIVIDUALIZED DEVELOPMENT

6.6.1 Introduction

As indicated earlier, a technical employee has the primary responsibility for her development. A person's supervisor has secondary, but nonetheless critical, responsibility as well. Two other classes of individuals can also play an important role in an employee's development: an employee's peers and staff members (for those employees who have a supervisory responsibility).

As more organizations embrace employee involvement principles, assessment by peers and staff members is becoming more prevalent. However, these two sources of feedback on strengths and developmental needs are still not utilized in the majority of organizations. This is very unfortunate, since peers and staff members both offer unique perspectives on areas needing improvement that an individual cannot gain through self-assessment or evaluation by his supervisor. By combining information from all of the various sources mentioned above (that is, self, supervisor, peers, and staff members), a technical manager can take advantage of a valuable technique that is known as "360-degree feedback" [8]. (In cases where this is relevant, customers can serve as an additional source of information in a 360-degree feedback program [8].)

6.6.2 Effectiveness Feedback Form

If a manager or nonmanagerial employee who works in a technical area wants to take advantage of feedback on his effectiveness from peers and staff members (if relevant), he can use the *effectiveness feedback form* (EFF) shown in Appendix E, Section E.1. The individual can request that the anonymous, typed feedback information be returned directly to him through interoffice mail or to a disinterested third party (for example, a member of the human resources staff). I recommend the use of a third party. Even though the written feedback is anonymous, some people may be more comfortable when they are able to give their candid comments to a neutral third party rather than directly to the person being evaluated.

If the third party has some expertise in development, she can also provide the individual being evaluated with her perspective on how to interpret and best utilize the information to improve on-the-job effectiveness.

It is critical that the individual requesting the feedback not try to determine who wrote certain comments. The information is provided anonymously, and the evaluators have a right to maintain their anonymity. The source of some comments may be identifiable because of the phraseology used or the individuality of the content. If these comments are negative, it is important that the individual being evaluated not confront or take any kind of action against the evaluator. It is essential to remember that the purpose of the activity is constructive. Therefore, all comments, even those that may have very negative connotations, should be considered to have been provided with good intentions (that is, to help the person being evaluated to improve his effectiveness).

The EFF requests favorable as well as unfavorable information. Although the primary purpose of the EFF is to provide information that can be useful for development, the feedback provided can also help a person understand or reinforce his understanding of work-related strengths. Because it requests both positive as well as negative information, the EFF presents more of a balanced picture of the individual being evaluated. This should make it more likely that the person being evaluated will perceive the exercise to be constructive rather than harmful.

The individual requesting the evaluation needs to determine how many people will complete the EFF. There is no magic number, but at least five should be involved to provide sufficient information and to help preserve the anonymity of the respondents. Most individuals probably will not need to request that more than ten people provide evaluations, but in certain situations a greater number may be helpful.

It is important that a person get a good cross section of evaluators in terms of anticipated favorability of comments (that is, not only those who are expected to give glowing evaluations should be included). Typically, only people who have worked with a person for at least six months and who have fairly regular contact with the person being rated should be included. It is not necessary to include everyone who works with the person being evaluated, just a representative sample of adequate size.

It is a good idea to include a person's supervisor among the evaluators as well. Even though most people already receive feedback on their work performance from their supervisors through formal and informal performance reviews, this provides another valuable perspective. In addition, sometimes supervisors provide more candid evaluations when they are anonymous.

At the same time that others are evaluating someone, the person might also do a self-evaluation using the EFF. He can then compare this to the other evaluations to see how consistent his perceptions are with those of coworkers.

6.6.3 Management Assessment and Development Inventory

As I indicated earlier, to get some basic developmental feedback from coworkers, a technical employee or manager can use the EFF herself. Alternatively, she can use the EFF with the assistance of an objective third party from within the organization. Another option is to use the services of a consulting psychologist or other outside professional who is well-qualified in the area of developmental counseling. Some such individuals may have access to a well-developed 360-degree feedback tool that is substantially more detailed than the EFF. The *Management Assessment and Development Inventory* (MADI) is such an instrument.

The MADI is an 87-item questionnaire that I developed, which involves rating a manager of technical employees on a five-point scale of effectiveness. In addition, the person who does the ratings can provide some written comments. There are eight general areas covered on the MADI. These are leading, motivating, communicating, delegating, decision making, planning, relating, and developing. Some sample items from the MADI are listed in Section E.2 of Appendix E.

A manager rates herself on the MADI and is rated anonymously by about ten coworkers. These usually include all of the staff members who report directly to the manager, several peers, and the manager's supervisor. Then I meet with the manager and review the feedback, using a standard format to present the results. A portion of a sample presentation of MADI results is indicated in Section E.3. (An actual report of results is typically about six pages long.)

At the meeting, I give the manager a copy of the MADI results and go over the highlights. These usually include the five to ten highest and lowest average ratings by coworkers, the three or four greatest differences between self-ratings and average coworker ratings, the three or four items that have the greatest range of coworker ratings, and the most significant favorable and unfavorable comments.

I also ask the manager to analyze all of the items and comments in more detail on her own. After this has been done, I ask her to write a "developmental action plan" that deals with the most significant areas needing improvement. I ask the manager to make the actions pragmatic, specific, and measurable, with implementation dates indicated. Several items from a sample developmental action plan are included in Section E.4 of Appendix E. (An actual developmental action plan is typically one to two pages long.)

I tell the manager to call me if she encounters any problems writing the action plan. Although I will provide assistance if it is requested, I want the individual to write it herself. This ensures the ownership that is critical to successful implementation; it is her plan, not mine.

After the manager has written the developmental action plan, we review it together. I try to ensure that it is realistic and that all of the key developmental needs are covered. I provide suggestions for modifications or additions as well, if necessary. Then I ask the manager to send a copy of the action plan to and discuss it with her supervisor.

The person's supervisor provides suggestions for changes or additions at their meeting, if necessary. The individual incorporates any changes into a finalized document, keeping the original and making copies for her supervisor and me. The person then implements the finalized plan. I suggest that the person's supervisor meet with the manager after an appropriate interval to give her feedback on progress or lack of progress.

One final suggestion that I usually add is that the person ideally should go through the 360-degree feedback process using the MADI again after 18 months or so. If the person has designed and implemented an appropriate action plan, the ratings given and comments made the second time should reflect improvements in the areas highlighted for development. In addition, repeating the process again gives the individual some new feedback that can aid in her further development.

6.6.4 Impact of the MADI and Similar Instruments

I have used the MADI, or other somewhat similar instruments, for over 15 years with hundreds of managers, including numerous managers of technical employees. The managers have been from all organizational levels, ranging from first-line supervisor to CEO. My experience has encompassed all types of firms, both manufacturing and service, in various industries. I have worked with top-level executives in Australia and the United Kingdom, as well as the United States. Although I have not yet used the MADI with managers in government, nonprofit organizations, or educational institutions, it is equally applicable in these contexts.

I use the MADI currently with a number of the client companies with which I work as a consulting psychologist. For example, Burdick, Inc., a very successful manufacturer of electrocardiograph equipment, is one of my clients. George Messina, Burdick's CEO, and some other members of Burdick management have asked me to use the MADI with a number of their managers. Some of these individuals have been people who manage technical employees.

Various authors have reported that using 360-degree feedback can be helpful in improving managers' abilities and performance [9, 10]. I have found that 360-degree feedback processes using the MADI or related instruments are the most powerful vehicles for improving managerial performance that I have encountered in my entire career. The impact of these programs is significant for several reasons. First, the feedback is highly individualized; each written presentation of results is unique to the particular person being rated. Second, the information received cannot simply be ignored because it is provided by people who work closely and regularly with the individual being rated. Third, the individual is expected to do more than just gain insight into his strengths and areas of developmental need. The person must design and implement a plan of action that requires *specific changes in behavior* in those areas where improvement is needed. Finally, the individual's direct supervisor is *closely involved* to ensure that these changes *actually occur.*

I have found that using 360-degree feedback typically results in *significant improvements in performance* as long as the participant diligently follows every step of the process. A few of the many successes that I have witnessed are now described.

When I was vice president of employee development for Sentry Insurance, I worked with a senior executive who received substantial feedback from others that indicated he tried to do everything himself. He agreed that he did this, and he wanted to change this ineffective behavior. He devised several specific strategies for delegating, including one that was particularly helpful to him. He simply made a tent card for his desk that said, "Can somebody else do this?" When he went through his mail every day, he would often receive memos that highlighted serious problems that needed to be handled at once. Through habit, he would begin to dictate a response to one of these memos, and then he would notice his reminder tent card. At this point, he would stop dictating, call one of his staff members, and ask her to come into his office. After providing adequate background on the problem, he would delegate to the staff member the tasks of responding to the memo and handling the problem. Subsequent discussions with the executive and his staff indicated that he showed tremendous improvement in his willingness to delegate to others rather than trying to do everything himself.

A manager of technical employees from one of Sentry Insurance's former subsidiaries with whom I worked received a very clear message from those who rated him that his staff meetings were lengthy, rambling, boring, and rather unproductive. He decided to ask his staff for suggestions on how to make the meetings more effective. He adopted all of the suggestions, and later comments from both him and his staff indicated that the quality of his meetings had dramatically improved. When he was again rated several years later, his average coworker rating on a question concerning how effectively he conducted meetings increased dramatically (that is, it went up by more than a full point, as compared to his initial rating). In addition, written comments indicated that conducting meetings was now one of his strengths, instead of one of the areas needing significant improvement.

6.6.5 Employee Assessment and Development Inventory

I developed another instrument, the *Employee Assessment and Development Inventory* (EADI), that can be used for the individualized development of technical employees who do not have supervisory responsibility. The EADI is somewhat similar to the MADI, but there are only 53 items and seven major areas relevant to nonmanagers. These areas are planning, leading, communicating, decision making, relating, doing, and cooperating. The steps in the EADI process (that is, rating, feedback, action plan development, and implementation) are identical to those in the MADI process. Some sample EADI items are indicated in Section E.5 of Appendix E.

I have used the EADI in my consulting work and have found that it is also a very effective tool for improving people's job performance, just as is the MADI.

6.7 SUMMARY

Although it may not be realistic to expect technical employees to change their personalities, they can usually change their behavior in certain work-related situations if they choose to do so. Although the primary responsibility for change/development rests with the individual, his immediate supervisor can and should also play a critical developmental role by serving as a coach and counselor.

Managers of technical employees can be of great help to their staff members in defining their developmental needs, identifying appropriate developmental vehicles, determining specific benefits before and after participating in a developmental experience, and providing feedback concerning developmental success.

Imitation learning can be a powerful tool that mangers can use to develop technical staff members' abilities. Ability development teams can be a valuable resource to supplement the coaching done by employees' immediate supervisors.

The 360-degree instruments, which provide individuals with feedback from peers, staff members, and supervisors, can be invaluable developmental resources. The EFF, MADI, and EADI are examples of such instruments. These tools can be used to highlight key developmental needs, which can then be addressed with specific action plans.

References

[1] Mone, Edward M., "Training Mangers To Be Developers," in *Careeer Growth and Human Resource Strategies: The Role of the Human Resources Professional in Employee Development*, Manuel London and Edward Mone (eds.), New York: Quorum, 1988.

[2] Kellogg, Marion S., *Closing the Performance Gap: Results-Centered Employee Development,* American Management Association, New York, 1967.

[3] Latham, Gary P., and Kenneth N. Wexley, *Increasing Productivity Through Performance Appraisal,* Reading, MA: Addison-Wesley, 1981.

[4] Sperry, Len, and Lee R. Hess, *Contact Counseling: Techniques for Developing People in Organizations,* Reading, MA: Addison-Wesley, 1974.

[5] de Board, Robert, *Counselling People at Work: An Introduction for Managers,* Aldershot, Hampshire., England: Gower, 1983.

[6] Klausmeier, Herbert J., and Richard E. Ripple, *Learning and Human Abilities: Educational Psychology,* Third Edition, New York: Harper & Row, 1971.

[7] Robinson, James C., "Behavior Modeling Training," in *Human Resources Management and Development Handbook,* William R. Tracy (ed.), New York: AMACOM, 1985.

[8] Ward, Peter, "A Turn for the Better," *People Management,* Vol. 3, February 9, 1995, pp. 20–22.

[9] Hazucha, Joy Fisher, Sarah A. Hezlett, and Robert J. Schneider, "The Impact of 360-Degree Feedback on Management Skills Development," *Human Resource Management,* Vol. 32, No. 2, Summer/Fall 1993, pp. 325–351.

[10] Carson, Mary Kay, "Subordinate Feedback May Foster Better Management," *The APA Monitor,* Vol. 26, No. 7, July 1995, pp. 30–31.

Chapter 7

Inhouse Training

7.1 INTRODUCTION

In addition to one-on-one development, which was discussed in the last chapter, employees can also be developed in groups using inhouse training. Many organizations provide inhouse courses for their technical employees. These are designed to improve their knowledge and job-related abilities, with the eventual goal of improving work performance.

7.2 HOW ADULTS LEARN

7.2.1 The Adult Learning Model

Some organizations have less effective group training programs than others because the model they use for their training approach does not take into account how adults learn. These organizations provide technical employees with training that is based on a model reflecting how children learn. That is, the model assumes that there is a subject matter expert (that is, the instructor), and this individual imparts knowledge to the trainees. This is done largely through lectures and other forms of one-way communication.

Many organizations use this model because it is one with which they are familiar. In other words, many of the people involved in this type of training were taught in this fashion when they were younger. As a result, they believe that this is the way that adults in the workplace should also be taught. Unfortunately, however, adults do not learn the same way as children do.

Malcolm Knowles developed an approach to adult learning that he called "androgogy"[1]. This approach is based on the idea that as a person grows, he changes. The individual becomes more interested in having greater self-direction in learning, being able to use his experience in the learning process, and having the ability to relate learning to life problems. This tendency, according to the model, tends to continue to increase until one reaches adulthood [1].

Many children will accept the one-way educational model, which focuses on the expert enlightening the student with her knowledge. However, most adults do not want to learn in this way, nor do they learn best in this fashion. Instead, adults prefer and probably learn best through the use of more interactive, participative methods. These approaches assume that the learner is not simply a passive "ball of clay" to be molded by the teacher. Instead, the adult learning model [1] recognizes that adults have had many diverse experiences in the past. These experiences can be shared among the members of a class, and trainees can learn from one another. Instructors can also learn from the trainees, in addition to the trainees learning from the instructors.

Adults do not like to learn via one-way lectures. When a lecturer goes beyond about 30 minutes or so, many adults find that their attention span tends to wane. As a result, to make learning maximally effective for adults, a much more participative, interactive approach needs to be used. In other words, a subject matter expert may present a brief amount of material (for example, about 15 or 20 minutes' worth). Then trainees can be asked questions about what was presented, and a two-way discussion can ensue.

Alternatively, after a brief lecture period, trainees can be asked to apply what was presented in some type of exercise. This can assess whether they, in fact, understood the material that was presented to them. An exercise can also assess whether an individual is able to apply what was learned to a practical work situation.

The instructor can review the trainees' performance in these exercises. If all of the trainees demonstrate that they understand and/or that they are able to apply the material that was presented, then the instructor knows that she can move on to the next topic. On the other hand, if a number of the members of the class demonstrate an inability to understand and/or apply the material, then the instructor needs to spend additional time presenting the subject matter. In this case, she might consider using a somewhat different approach.

If an instructor uses exercises that require applying what was learned and interactive discussions, she would be following the principles of androgogy.

7.2.2 Sensory Modality Preferences

Instructors need to take into account how trainees prefer to learn, sometimes referred to as "sensory modality preferences" [2]. Some people prefer to learn primarily through a visual approach (for example, by reading material). Other individuals may prefer to learn by hearing (for example, by listening to a lecture). Still other people may prefer to learn through kinesthetic means (for example, some type of hands-on exercise). An instructor of technical employees can take into account the various sensory modality preferences within a group by asking people prior to the class which of the three modalities is their favorite. With this information, the instructor can then ensure that material is presented using each of the relevant modalities, hence ensuring that the auditory, visual, and kinesthetic learners will all benefit from the instruction.

7.2.3 Cognitive Styles

Instructors of adults also need to take into account the various ways that different trainees learn. These are sometimes referred to as "cognitive styles" [2]. One style, for example, is known as "impulsivity versus reflection" [2]. This style can be viewed as a continuum, with one end representing the extremely impulsive individual (that is, someone who processes information very rapidly and makes "snap" decisions) and the other end the extremely reflective individual (that is, one who tends to analyze information carefully and thoroughly and to evaluate alternatives before coming to a conclusion). Although many technical employees probably operate more toward the reflective end of the reflection/impulsivity continuum, instructors will find impulsive people in their classes as well.

An instructor can determine whether the class participants are primarily impulsive or reflective learners by asking them questions prior to the class. For example, one question might be, "Do you often make quick decisions?" Another example of a relevant question might be, "Do you often thoroughly analyze a problem and evaluate all of the various alternatives before determining the solution?"

Another approach that an instructor could use is to complete some type of questionnaire that measures impulsivity versus reflection. For example, a brief questionnaire that assesses this is included in an article that appeared in *Industry Week* in 1976 [3].

It is important for an instructor of a group of technical employees to present material in a fashion that is appropriate for both reflective and impulsive learners, if both types exist in the class. For example, an instructor might ask students to brainstorm and indicate, without evaluation, some possible ways to deal with a particular situation. An instructor might also ask the students to analyze and to evaluate a particular approach for dealing with a problem. The former exercise might be best suited for impulsive learners, while the second exercise might be best suited for learners of the more reflective type.

7.3 MULTIPLE-SOURCE TRAINING NEEDS

It is important that an instructor ensure that some type of training needs analysis is done prior to conducting classes. A training needs analysis is critical because it helps to identify the highest priority training needs that exist. These high-priority needs, in turn, can be translated into the most important topics to be presented. In addition, a training needs analysis can provide an instructor with some detailed information concerning what aspects of a topic are most important.

For example, a manufacturer may do a training needs analysis and find that its manufacturing engineers need to be aware of the latest state-of-the-art equipment for manufacturing its products. This topic, then, might be an important one to cover in a training session. Getting more specific information concerning the types of equipment that the manufacturing engineers are most interested in learning about would be even more helpful in designing an appropriate training session.

Training needs may be assessed in various ways. One approach is the use of questionnaires [4]. Advantages of questionnaires include a relatively low cost and that a great deal of information can typically be collected in a fairly short time [5]. I have used needs analysis questionnaires in various organizations and have found them to be very useful.

There are various formats for training needs analysis questionnaires. One format that could be very helpful to trainers is one that my department used when I worked at Sentry Insurance. It involves listing various topics that might be covered in training sessions and asking individuals to rate each item in terms of its importance and their current knowledge in that area. Both current knowledge and importance can be evaluated on a five-point scale. Regarding importance of a topic, this can range from least important (1) to most important (5). Regarding current knowledge of a topic, this could range from very low (5) to very high (1).

In ranking the various topics in terms of both current knowledge and importance, the two scores can simply be multiplied together. The resulting product can then be ordered by rank, starting with the greatest product and going to the smallest product, in decreasing order.

Using this approach, the highest possible rating of a topic in a needs analysis questionnaire would be 25 (that is, very high importance and very low current knowledge). Conversely, the lowest possible rating would be 1 (that is, very high current knowledge and very low importance). Using this approach, an organization can ensure that its training programs address the highest priority training needs.

One additional complicating factor that needs to be addressed, however, concerns who completes training needs analysis questionnaires. They can be completed by current technical employees, by their immediate supervisors, or by the top management of a technical area. Each of these groups of individuals tends to view training needs from a different perspective. As a result, the findings of a training needs analysis conducted with each of these three groups might well differ. In fact, in most situations, although there may be some similarities, there undoubtedly will be some major differences in the way that each of the groups perceives the high-priority training needs.

Each of the three groups that can be surveyed concerning training needs has a unique perspective. Each has some distinct advantages over the other two, as well as some disadvantages. For example, in my experience, incumbents are the closest to a particular position, since they are doing it. As a result, they have the capacity for making some of the most accurate observations concerning a training needs analysis. However, incumbents also sometimes tend to identify training "wants" in addition to training "needs." In other words, incumbent technical employees sometimes demonstrate a bias in a training needs analysis questionnaire. They tend to choose topics about which they would like to learn more as opposed to topics about which they need to learn more to do their jobs more effectively.

Each of the three groups mentioned previously has both advantages and disadvantages regarding its perspective on training needs. Thus, it seems to make the most sense to use a combined source training needs analysis questionnaire. In other words, the top management of a corporation and/or a technical area can complete a training needs analysis questionnaire. In addition, the direct supervisors of technical employees can complete one. Finally, technical employees themselves can also respond to a needs analysis questionnaire. These three sources of information, then, can be combined to highlight the most important topics for training programs.

This combined approach can be done in a number of different ways. One way would be simply to compute an arithmetic mean of the perceptions of each of the three groups concerning the most important training needs. Another approach that might yield somewhat similar but not identical results would involve selecting the top two or three training needs as identified by each of the three groups.

Regardless of how the information is combined, a multiple-source training needs analysis questionnaire is probably superior to any single-source training needs analysis questionnaire. In addition, a multiple-source approach is undoubtedly significantly superior to having no formal means for assessing training needs.

7.4 TRAINING THAT REALLY WORKS

There are certain elements that can greatly enhance the value of technical training that is done in an organization. It has been my observation that in most organizations, the weakest link in a training program concerns what happens after the program is completed. A similar observation was made by Silberman [6]. I have observed many training programs that were well conceived and based upon an effective training needs analysis. The programs were well executed, and participants learned a great deal. In addition, participants in nearly all of the training programs that I have witnessed left the training session with very good intentions. Unfortunately, these good intentions often are not translated into action. People go back to their jobs and typically immediately get inundated with various problems and projects. As a result, people often find that they do not apply what was learned to any appreciable extent.

If this situation of not being able to translate good intentions into action is a frequent scenario, what can be done to change this? Various aids can be used in this regard [6]. One fairly simple technique that can be used is to put the key points covered in a training session on a small pocket-size "course reinforcement card." Such a card can be easily carried with an individual and referred to frequently to ensure that he applies what was learned in a training session to the job.

Another technique that can be very valuable in ensuring that what is learned in a training session is used on the job can be termed an "action plan" [6]. This approach involves doing something at the end of a training session that typically is not done: making up a plan that highlights exactly how an individual is going to translate what was learned into improved on-the-job behavior.

Trainees may have some difficulty initially in coming up with an action plan. This is probably due to the fact that few courses require doing this, so most people have not had experience in designing a developmental action plan. However, if a course instructor provides some guidance and perhaps some examples of what might be included in an action plan, trainees can use this approach to improve their on-the-job effectiveness.

To illustrate what might be included in an action plan, I offer the example of a manufacturing engineer who has just completed an inhouse training course on the use of robotics. After completing such a training course, this individual might design an action plan that includes such statements as (1) to recommend no less than three manufacturing process improvements involving robotics within the next 12 months, and (2) to implement at least one manufacturing process improvement involving robotics within the next 12 months.

The example mentioned above illustrates what is important in an action plan: some very specific, measurable statements that indicate precisely how a learner will be using the knowledge gained from a course on the job. When such developmental action plans are used and shared with a technical employee's supervisor, the probability of the individual actually using what was learned on the job in the near future is extremely high. Conversely, when such action plans are not used, a technical employee's "good intentions" concerning using what was learned in a course are often not translated into action.

One final point concerning the design of training that truly works is that a trainee's supervisor must be closely involved in the training process. The results of one study indicated that an often-cited reason for training failures is a lack of managerial support [4]. Unfortunately, in many organizations the following scenario takes place all too often. An employee goes to his manager and indicates an interest in taking a particular internal course. After the course is completed, the supervisor asks the employee how the course went, and the employee says "fine."

The supervisor then adds something, such as, "Well, that's good." This constitutes the extent of the supervisor's involvement in the training process.

If a technical manager truly wants an employee to get the maximum benefit from an internal course, he cannot repeat the above scenario. Instead, a technical manager needs to demonstrate a strong interest in the training process right from the beginning [6, 7]. First, a manager should work closely with an employee in defining her developmental needs as indicated in Chapter 6. If one of these developmental needs can be met by an internal training course, then a technical manager should strongly urge an employee to take this course. He should also remind the employee about taking the course, should the employee forget to do this.

Prior to taking an internal course, a technical manager should sit down with an employee and discuss specifically the various knowledge and skills that the employee should gain from the course. He should also talk about how the knowledge and skills can be applied on the job. After an employee has taken a class, a technical manager should review what was discussed by him and the employee prior to the class. The manager should also review any specific developmental action plans with the employee. The manager should then ask how he can assist the employee in using what was learned in the course on the job.

After a technical manager has given an employee sufficient time to implement what was learned in the course, the manager should meet with the employee. At this time, the manager can determine if, in fact, the employee has applied what was learned in the course on the job. If the employee has demonstrated that she has applied the knowledge from the course in her position, the technical manager should provide some positive reinforcement concerning the employee's success.

7.5 EVALUATING TRAINING EFFECTIVENESS

Don Kirkpatrick wrote an article quite a long time ago on evaluating training effectiveness that is just as valid today as when it was written. In the article, Kirkpatrick pointed out four basic ways in which training can be evaluated: by (1) measuring participants' reactions to the training, (2) measuring what was learned in the training program, (3) evaluating behavior change, and (4) assessing a change in results following the training program [8].

Unfortunately, the most prevalent form of training program evaluation is still measuring participant's reactions. Although finding out whether a trainee enjoyed a particular program is not unimportant, it is probably the weakest indication of a training program's effectiveness.

Often, a course that a trainee enjoys greatly is not one that provides a great deal of information that can be applied on the job. Although speeches are not training programs, they can serve as an example to illustrate my point. I have heard many speeches that I have found to be extremely interesting and that I enjoyed greatly. I liked them because the speaker had an interesting style that maintained my attention. Unfortunately, however, if I asked myself what I learned from some of these speeches, I would have to say that my response in most cases would be "very little."

This also applies to training programs. A trainee may enjoy a program greatly but may learn very little that can be used on the job [4].

The second primary way of evaluating a training program's effectiveness is to measure the learning that has occurred. Probably the most common means of doing this is to give a trainee some type of test at the end of a course. If a test is well designed and a good measure of what was supposed to have been learned in a course, the test results can provide useful information. Obviously, before someone can use some specific knowledge or skills on the job, he first must possess this knowledge or these skills. If someone has certain knowledge, he may not apply it. However, if he does not possess the knowledge, there is no way that it can be applied.

Of the various ways to measure training program effectiveness, evaluating what was learned is probably the second most prevalent technique that is used in organizations. It is helpful to know whether a trainee has gained certain knowledge, just as it is beneficial to determine whether she enjoyed a course. This by itself, however, is still not a particularly good way to measure training effectiveness because, although someone has learned something, this does not necessarily mean that she can or will apply it on the job.

The third means of evaluating training program effectiveness is to assess behavior change. In my estimation, this is a much better means of evaluating a training program's effectiveness than the first two approaches that were discussed above. Unfortunately, however, assessing behavior change is usually more difficult than determining reactions and learning.

Despite some possible difficulties, however, behavior change can be assessed. There are a number of possible sources that can be used to determine how much behavior has changed as a result of having completed a course. First, a trainee himself can be questioned concerning behavior change. A rather open-ended question can be used, such as, "How has your behavior changed since you have completed the course?" Depending on what a person says in response to this question, additional questions can be asked.

Another approach is to design some specific questions that attempt to assess behavior change. For example, suppose a consulting engineer took a course in improving writing skills. The employees' supervisor might ask her, since taking the course:

1. How has your writing of reports for clients changed?
2. How has the grammar in your reports been modified?
3. How has the punctuation in your reports changed?
4. What changes in the quality of your report writing have you noticed?

5. What differences have you noted regarding the time it takes to write reports for clients?
6. What changes in clients' comments concerning your reports have you noticed?

Another source that can be used to assess behavioral change in a trainee is the individual's immediate supervisor. A technical manager can ask himself questions similar to those previously indicated for a trainee. For example, the manager of the consulting engineer referred to in the above example might ask himself, "How has this employee's writing changed since she has completed the course in improving writing skills?"

Two other possible sources concerning the evaluation of a trainee's behavior change upon the completion of a course are the trainee's peers and, if relevant, her clients. Questions such as those indicated previously can be used with these individuals to assess behavior change in an individual following the completion of a course.

The last means of evaluating training effectiveness is the determination of changes in results. Of the various means of evaluating training effectiveness, this is, in my opinion, the most significant indicator of training effectiveness. However, typically this is also the most difficult to measure (that is, in terms of determining the specific ways in which taking a course has led to changes in organizational results).

Organizational results can be measured in a variety of ways. Using the previously mentioned example of the consulting engineer who took a course in improving her writing skills, one possible measure of organizational results might be client retention. In other words, the consulting engineer's immediate supervisor might choose to evaluate whether his employee's retention of clients has increased since taking the course in improving her writing skills. The assumption here might be that if the consulting engineer is able to write better reports after taking the course, she should have more satisfied clients. One measure of the degree of client satisfaction might be some type of indication of client retention, such as the percentage of clients who choose to do future business with this particular consulting engineer.

Using the same example, another change in results that might be a bit less indirect could be clients' evaluation of the consulting engineer's work, including her reports. Suppose the consulting engineer's supervisor had some type of measure of client satisfaction prior to the consulting engineer taking the writing skills course. This measure, if quantified (for example, in the form a numerical rating), could be compared to the client satisfaction rating sometime after the individual completed the writing skills course.

One of the problems associated with using results as an indicator of training program effectiveness is that there are numerous factors that influence results. As a result, it is often very difficult to say with any degree of certainty that taking a particular course influenced the organizational results [4]. For example, in the previous example, the client retention rate of the consulting engineer could be impacted by numerous factors. Only one of these is the writing ability of the consulting engineer. If the consulting engineer's client retention rate increased some time after she took the writing skills course, this might be due, at least in part, to taking the course. However, the increase in the client retention rate might also be due to the fact that the consulting engineer resolved a number of personal problems over the past six months or so and, as a result, was able to devote more time to and focus more attention on her clients' concerns.

This problem is diminished to some extent if a less indirect results measure is used, but it is not eliminated. For example, suppose a survey of client satisfaction is completed sometime after the consulting engineer has completed the writing course. Suppose also that the survey results indicate an improvement in satisfaction ratings. This might be due, at least in part, to taking the course. However, it could also be due primarily or exclusively to other factors. For example, the clients surveyed after the course was taken might be more lenient in their ratings than those surveyed previously. As a result, what appears to be an increase in client satisfaction may, in fact, reflect only a change in the rating standards between the first and the second groups of clients.

As can be seen from the previous discussion of the various ways of evaluating training effectiveness, each of the various means has advantages and disadvantages. As a result, perhaps an ideal approach is for an organization to employ each of the four techniques [4]. This approach enables an organization to experience the advantages of each evaluation method while minimizing the disadvantages to some extent, as a result of using multiple measures.

7.6 TYPES OF INHOUSE TRAINING

7.6.1 Technical Training

Obviously, one important type of inhouse training that can be provided to technical employees is technical training. This can be a very cost-effective approach, particularly if the instructors of the various courses are internal employees. Many employees enjoy teaching and sharing their knowledge with others. As a result, many people are quite willing to teach classes regarding technical matters without being compensated above and beyond their regular compensation.

Even if a company feels that it is appropriate and/or necessary to pay inhouse instructors additional money for their teaching duties, this is often a very good investment. The instructors know the company very well and, as a result, are often able to tailor the material to the specific needs of the organization.

If a company's management feels it is necessary or most appropriate to use outside instructors, this is often more costly. In addition, the teachers typically find it more difficult to tailor the course content to the specific organizational needs since they are not employed by the company [9]. On the other hand, one major advantage of using outside instructors is that an organization, if it spends sufficient time selecting instructors, can ensure that they are both extremely knowledgeable regarding the subject and excellent presenters. Unfortunately, there are many people, both external and internal, who may be very knowledgeable about a subject but not particularly effective in conveying this information to others. An organization may well have some individuals who are extremely knowledgeable in various subjects as well as outstanding teachers. However, in many cases the ideal instructor may not be employed by an organization.

7.6.2 Orientation/Teaching About the Company Culture

Another key type of internal training of technical employees involves the orientation of new employees. Many organizations have no formal orientation programs for new employees, while some have programs that are quite structured and formal. Obviously, an organization can rely on its technical managers to orient new employees properly. However, what if the organization does not provide all managers with a very structured outline or lesson plan regarding the orientation session? In this case, there is often great diversity, both in terms of what is presented and how effectively it is taught.

To avoid this problem, I recommend that an organization develop a rather formal, structured program. To ensure the greatest consistency in terms of content and presentation style, one or several individuals can be selected to orient all new technical employees. Alternatively, various technical managers can simply be responsible for orienting the individuals whom they supervise. This approach might not result in the same degree of consistency. However, exemplary results can be obtained with the use of various internal managers as long the program is structured in a fairly detailed fashion, and monitored to be sure that people follow the "organizational lesson plan."

One facet of orientation programs for technical employees that I have found to be relatively rare in my experience is content that deals with the organization's specific "culture." Omitting this topic from an orientation program is, in my estimation, a major error. This view is supported by others [10]. By doing so, an organization forces new employees to try to ascertain the various attitudes, beliefs, and practices found in the organization on their own. Such an approach results in many employees not fully understanding all aspects of the corporate culture for quite some time. In addition, some aspects of the culture may not be picked up at all by certain individuals.

For example, one organization might have, as part of its culture, an unwritten practice that people are expected to work extremely long hours to do their jobs most effectively and to enhance overall organizational performance. Many individuals probably would pick this up very rapidly, even if they are not provided with this information in a formal orientation program. Others, however, might not realize this immediately, or perhaps even after quite some time, without being formally informed about this process. For example, most technical employees might routinely come in on Saturday mornings. Suppose a new employee is not specifically told about this. Suppose also that he does not hear any comments about this practice, and the employee did not work on Saturdays in previous positions. In such a case, he might not realize for quite some time that the organizational practice exists.

I can provide another example of one aspect of corporate culture that might be best presented in a formalized program. Suppose that technical employees are expected to cooperate, not compete, with technical employees in other parts of the company. If this expectation is not presented in a formal orientation program, a technical employee might not become aware of this for a period of time. This might be particularly true if the employee was one who previously worked for an organization where cooperation was the exception rather than the rule.

Naturally, if a particular aspect of an organization's culture is presented in a formal orientation program, this must actually be demonstrated in the organization as well as discussed. For example, the cooperation versus competition referred to previously would need to be consistently demonstrated by numerous employees for a new employee to gain a true understanding and acceptance of the existence of this practice. For this reason, it is important that the various aspects of an organization's culture that are presented in a formalized orientation program be agreed upon by the employees at all levels. If this does not occur, upper management and/or the developers of an orientation program may tend to describe the organization's culture in terms of what they would like it to be, as opposed to what it really is. There is nothing wrong with describing the culture that upper management ultimately would like to see in an organization, as long as this is explained in the orientation session. Problems occur, however, when new employees are told that an organization's culture is described in a certain way, when it does not truly operate in this fashion.

For example, suppose an organization told its new employees that individuals are expected to cooperate with people in other departments rather than compete with them. Suppose also that this does not actually happen in practice. The result is likely to be a credibility gap with new employees. In other words, this situation may cause new employees not to believe what they are told by the organization.

7.6.3 Ability Training

One final type of training that can be conducted internally for technical employees might be called "ability training." I use this term to refer to training that does not relate to the technical content of a job but is nonetheless potentially helpful to technical employees. Examples of such training might include, for example, developing listening abilities, improving writing abilities, enhancing interpersonal abilities, and strengthening decision-making abilities.

When I worked for Sentry Insurance, the company offered various ability training courses. These were very popular, and they were very helpful to participants in improving their job performance. The instructors of these courses at Sentry were various internal employees who were identified as being good instructors and knowledgeable in the subjects which they taught.

As I indicated in the section on technical training, there are advantages as well as disadvantages of using internal employees to teach inhouse classes. One approach that seems to be ideal, in my opinion, is to try to find inhouse people who can teach the various ability training courses effectively and are knowledgeable regarding the content of these courses. If such individuals can be identified, they can be used as instructors, assuming they are willing to teach. If no qualified internal candidates can be found or no qualified internal candidates are willing to teach certain courses, outside individuals can be identified to teach these courses.

7.7 SUMMARY

Organizations that provide effective inhouse training for technical employees use an appropriate educational model that emphasizes interactive learning. They also recognize sensory modality preferences and cognitive style differences.

Training needs analyses that involve multiple sources of input (that is, from trainees, supervisors, and top management) are very effective. Other aspects of inhouse training that work well include the use of course reminder cards, developmental action plans, and substantial supervisory involvement in the inhouse training process.

Inhouse training can be evaluated based on participants' reactions, learning, behavior change, or results. Each method of evaluation has advantages and disadvantages. An approach employing multiple evaluation methods seems advisable. Some types of inhouse training include technical training, training regarding the organizational culture, and ability training. All three types of training can be very valuable to an organization.

References

[1] Pont, Tony, *Developing Effective Training Skills*, London: McGraw-Hill, 1991.

[2] Kogan, Nathan, "Educational Implications of Cognitive Styles," in *Psychology and Educational Practice*, Gerald S. Lesser (ed.), Glenview, IL: Scott, Foresman, 1971.

[3] Soat, Douglas M., "How Impulsive Are You?" *Industry Week*, March 8, 1976, pp. 32–33.

[4] Sims, Ronald, *An Experimental Learning Approach to Employee Training Systems*, Westport, CT: Quorum, 1990.

[5] Goldstein, Irwin L., Eric P. Braverman, and Harold L. Goldstein, "Needs Assessment," in *Developing Human Resources*, Kenneth N. Wexle, (ed.), Bureau of National Affairs, Washington, D.C., 1991, pp. 5-35–5-72.

[6] Silberman, Mel, *Active Training*, New York: Lexington, 1990.

[7] Saint, Avice, *Learning at Work*, Chicago: Nelson-Hall, 1974.

[8] Kirkpatrick, Donald L., "Evaluation of Training," in *Training and Development Handbook*, R. L. Craig and L. R. Bittel (eds.), New York: McGraw-Hill, 1967, pp. 87–112.

[9] Tracey, Willliam, *Designing Training and Development Systems*, New York: AMACOM, 1992.

[10] Trice, Harrison and Janice Beyer, *The Culture of Work Organizations*, Englewood Cliffs, NJ: Prentice-Hall, 1992.

Chapter 8

Other Means for Developing Technical Employees

8.1 INTRODUCTION

The last two chapters discussed one-on-one development and inhouse training. This chapter deals with some additional means for developing employees, including outside courses, formal reading programs, career development, succession planning, and team building.

8.2 OUTSIDE COURSES

All development that technical employees need cannot be done inhouse. Even if a technical manager is extremely knowledgeable, she ordinarily cannot provide all of the development that her technical employees need. In addition, even if there are other technical employees on the staff who are very experienced, they ordinarily will not be able to provide all the development that the rest of the employees need. As a result, some employees will need to take outside courses to meet certain developmental needs.

As mentioned in Chapter 6, a technical manager plays a key role in helping an employee to define his developmental needs. In addition, as was also mentioned earlier, a technical manager is often aware of certain outside courses that are unknown to employees. Thus, a technical manager can be very helpful to an employee in identifying appropriate outside courses that are relevant to the employee's major developmental needs. However, an employee himself needs to take the primary responsibility for identifying appropriate outside courses.

Technical employees are often familiar with certain external educational offerings. However, how can an individual increase her awareness of relevant outside courses? First, an employee can mention to her supervisor as well as other employees in the department that she is looking for a course on a particular topic. Many educational organizations (for example, universities, university extension institutes, technical colleges, and professional organizations) send out mailings. Even though a technical employee may not receive certain mailings, someone else in the department may.

In addition, a technical employee might contact a relevant professional association regarding course offerings. These professional associations could be rather large groups covering a broad range of positions and specialties (for example, a professional organization that has electrical engineers as its members) or a much smaller, more specific group (for example, an association that deals with fiber optics).

Another option is to contact the universities in the area regarding possible relevant course offerings. Alternatively, a technical employee may be aware of one particular university that may not be in the area that specializes in a particular technology, or of an individual faculty member at a particular school who is an expert in a particular area. The school and/or faculty member specializing in a particular area of technology can then be contacted and asked about relevant course offerings in the subject area.

There are many courses offered in a multitude of subjects. Some are very good, some are average, and some are not very good. To attempt to find out whether a particular course is relevant to a technical employee's needs and whether it is effectively presented, an individual might ask a sponsoring organization for the names of previous course attendees. These individuals could then be contacted and asked about both the content and the effectiveness of the presentation.

Another way to learn more about a particular subject area is to ask about audio tapes and/or materials from a previous presentation. This might be particularly helpful when a course will not be offered again for quite some time, but a technical employee needs to become familiar with the subject matter right away.

Another option that a technical employee might take if a relevant course is not offered in the near future is to ask a subject matter expert if he would be willing to conduct another course in the near future. If this is not feasible, a technical employee might ask a subject matter expert if he would be willing to conduct a one-on-one session. Some subject matter experts might be willing to do this for a fee, and others might even be willing to spend some time with a technical employee without charge.

8.3 FORMAL READING PROGRAMS

Another very helpful developmental technique that does not seem to be used very often is a formalized reading program. By using this approach, a technical manager might identify a bibliography of various publications (for example, books, pamphlets, articles in professional journals) that deal with a particular subject area. Technical employees who need to increase their knowledge in a particular area could then be asked to read the various publications. Some type of formalized schedule, including deadline dates, could be developed regarding the various publications. In addition, a technical manager could set up a schedule of meetings with a technical employee. At these meetings, a manager could ask the employee various questions to determine if he has read the material and understands it. In addition, a technical manager could provide additional information at these meetings to supplement the reading material and to help to expand the employee's knowledge in the area.

If a technical manager wanted to formalize the reading program even further, she could devise tests that measure an employee's understanding of the subject matter. A technical manager could review these tests with an employee after they are taken to help correct any misunderstanding. In some cases, a manager might be able to find relevant tests that are developed and perhaps even administered by third parties (for example, relevant individuals in certain professional organizations).

Another option that a technical manager might follow is to ask another employee in the department who is very knowledgeable regarding the subject to meet with the technical employee after she has read a particular article or a series of publications. Such meetings, although they would be primarily of benefit to the employee who is the "nonexpert," could also be of significant benefit to the other employee who is knowledgeable concerning the particular subject. Preparing for a discussion sometimes forces an individual to do some preparation and learn some new material that he would not ordinarily have done. In addition, the nonexpert might have some questions during the session that might stimulate the other employee's thinking and perhaps even cause her to do some further reading on the subject that she might not otherwise have done.

If several technical employees are involved in a formal reading program on a particular subject at the same time, a technical manager can develop a small class that is built around the reading material.

8.4 CAREER DEVELOPMENT

Most technical employees are interested in developing their technical skills concerning their current job. In addition, however, many employees are also interested in developing knowledge and skills that can help them in possible future positions [1].

Unfortunately, most technical managers do not focus as much of their time and attention on dealing with technical employees' career development as employees would like. I have found significant evidence to support the previous point in my work with the Management Assessment Development Inventory (MADI), which was discussed in depth in Chapter 6.

One of the sections on the MADI is concerned with "developing." There are various questions concerning a manager's development of others, many of which deal with the topic of career development. About 80 percent of the technical managers who have been involved with the MADI program thus far were rated lower in the developing category than in any of the others. This indicates to me that most technical people feel that their managers simply do not devote as much time to employees' career development as they would like. This is unfortunate, because employees should feel that their managers are concerned about their career development and their welfare in general.

Why won't most technical managers deal with the topic of career development to the extent preferred by employees? There are probably a number of different reasons. For example, one reason might be that some technical managers feel somewhat uncomfortable or ill-equipped to deal with the topic of career development. Another reason might be that career development is a topic that involves many unknowns, which may cause some technical managers to feel somewhat uncomfortable.

However, I believe that the major reason why most technical managers do not discuss career development as much as employees would like is that developing employees for possible future positions is not a top priority for technical managers in most organizations. Technical managers are paid to get results today, not to prepare employees for possible future positions. I am familiar with very few organizations that consider developing employees as a major factor affecting a technical manager's salary and/or bonus.

In addition, besides not being positively reinforced for spending time on the subject of career development, there are actually significant drawbacks associated with spending time preparing an employee for another possible future position. If a technical manager helps to prepare an employee for another job outside of his department and the employee gets it, the technical manager is penalized by losing a good employee.

Considering the various information that was just discussed, a technical manager may ask himself, "Why should I spend time on career development?" The answer is that a technical manager needs to consider the issue of career development from the point of view of the organization as opposed to from his individual or departmental perspective. It is true that by helping a technical employee prepare for another possible future job, a manager may lose that person in his department. Some such employees may even leave the organization. However, the overall benefits of preparing employees for possible future positions greatly outweigh the potential drawbacks, at least from an organizational standpoint. If all technical managers spend a sufficient amount of time developing people for future career opportunities, this greatly increases the probability that when such an opportunity does arise, someone in the organization will be prepared to do the job effectively.

I have witnessed numerous organizations in which very little time is spent on the topic of career development. In such organizations, when a key opening occurs, there is typically no internal person who is qualified to do the job. As a result, outside candidates need to be considered for such openings.

It is not necessarily undesirable to fill certain openings with external candidates. In fact, I feel that it is usually desirable to fill about 20 percent of the openings with external candidates because doing so helps to "infuse new blood" into the organization. This tends to provide new ideas and to reduce provincial thinking to some extent. However, when 80 percent or more of the key openings that occur are filled from the outside, the advantages of having external candidates are outweighed by the disadvantages. When most positions are filled from the outside, internal employees begin to feel that they are being ignored and considered to be unimportant by management. Such feelings tend to erode technical employees' motivation. Employees who are very advancement-oriented and mobile often tend to leave organizations that do not provide adequate internal career opportunities. When such individuals leave, it is usually quite obvious to management concerning the reason for their departure. If enough good employees leave the organization because of a lack of career opportunities, the management in most organizations will begin to address the issue of career development [1].

However, in many organizations that I have observed, people do not leave an organization when they are passed over for promotions. Often, such individuals are very "locked into" the community in which they and their families live. As a result, such individuals frequently do not quit but do something that, in some cases, is much worse: they stay. Such individuals remain with an organization and, in some cases, simply "retire in place." These technical employees do not stop doing their job, because they realize that this would eventually result in their dismissal. Instead, they, consciously or unconsciously, decide to do just enough to get by. They perform the minimum requirements of their position, but do not put in the extra effort that is typically expended by highly motivated employees. In addition, these retired-in-place employees often stop providing new ideas and lose interest in developing and improving their skills in their current jobs.

This retirement-in-place phenomenon can be very insidious in an organization because the management in most organizations does not realize that it is occurring. Although it is difficult to recognize and measure, I am nonetheless convinced that it is very real, because I have seen this phenomenon operating in certain organizations. The loss of productivity that results from this phenomenon can have a significant adverse effect upon the overall performance of an organization. In addition, the divisiveness that frequently results from bringing in numerous new employees in key positions from the outside can also be a very disruptive factor that adversely affects an organization's performance.

Having discussed the various reasons why career development is important and the potential dangers of not dealing with this issue, I want to discuss suggestions concerning how a technical manager can deal with this topic. First, I strongly recommend that all managers of technical employees meet with each employee at least annually to discuss her career development. Some technical managers might feel that spending a great deal of time on the topic of career development is wasted time since employees, in general, seem to have very little organizational loyalty anymore. It is true that most employees no longer have the same degree of allegiance to their organizations that they may have had 15 or 20 years ago. Undoubtedly, this is, to a great extent, the result of organizations demonstrating significantly less loyalty to employees in recent years, as evidenced by the ubiquitous "downsizing." As a result of this diminished loyalty on the part of both employers and employees today, however, I would argue that this is all the more reason for technical managers to devote sufficient time to the career development of their employees. Doing so may provide at least one piece of evidence that the manager and the organization "care" about the employee more than the "typical" managers and organizations do.

When a manager of technical employees meets with his people to discuss their career development, he needs to prepare adequately for such meetings [2]. Preparation involves thinking about:

1. What are this employee's key developmental needs with regard to her current job?
2. What are some positions within the organization in which the employee might be interested?
3. What are some positions for which the employee might be qualified?
4. In addition to the developmental needs mentioned previously, what are some additional needs that the employee has with regard to possible future positions?
5. How might these developmental needs be met?

Managers of technical employees who are already quite busy may be reluctant to spend time on another activity: the career development of his employees. However, the time spent on this issue is well worth it in terms of the likely increase in employees' motivation, commitment, effort, and willingness to remain with the organization [3].

8.5 SUCCESSION PLANNING

Career development focuses on the individual technical employee in terms of her career interests. The complement to career development, from an organizational standpoint, is succession planning. Succession planning deals with ensuring that replacements are available for key positions when needed. Although the focus of succession planning is the organization's needs, an individual's interests cannot be ignored [3]. Focusing on an organization's needs without taking into account the career interests of a technical employee leads to problems such as the following. An individual is slated as a likely replacement candidate for a key position, but no one bothers to verify with the individual that he is actually interested in this career opportunity. When the incumbent in this key position leaves the organization, the position is offered to the technical employee who was slated as a replacement, and this individual turns down the job. Members of management are shocked, and they must scurry in an attempt to find a replacement for the individual who vacated the key position.

Recognizing the importance of the combination of career development with succession planning, how does a technical manager do an effective job of succession planning? First, a manager of technical employees needs to develop an organizational chart of her unit. After having done the above, a technical manager needs to develop a chart, such as the one indicated in Figure 8.1. The chart identifies incumbents for each key position, potential successors for the incumbents, and an indication of approximately when each individual is expected to be ready to assume the position.

As is illustrated in Figure 8.1, individuals' readiness is often described in terms such as "ready now," "ready within one or two years," or "ready in more than two years." Obviously, a technical manager's determination of this readiness is not a precise science. All that a technical manager can do is to give her best esti-

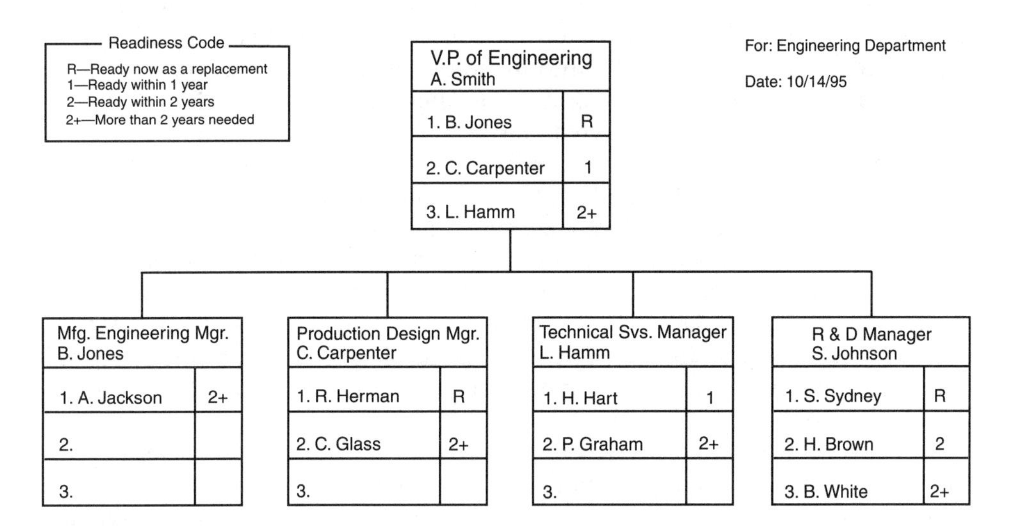

Figure 8.1 Succession-planning chart (fictitious example).

mate as to when each individual listed as a potential replacement for each position might be ready to assume that slot.

I have had fairly extensive experience in assisting organizations with their succession-planning processes. Through trial and error, I discovered a number of key ingredients of a successful succession-planning process. First, as mentioned previously, succession planning must be done in concert with career development. Probably the most important issue that must be addressed under the career development topic is the identification of the potential replacements' key developmental needs and *the implementation of actions that meet these needs*.

A major failure in the succession-planning process occurs when nothing is done to meet individuals' developmental needs with regard to possible future positions [4]. When such implementation does not occur, a scenario such as the following inevitably occurs. An individual is listed as a potential replacement for a particular position but is indicated as not being ready for two years. Since nothing specific is done within the next year to address the individual's specific developmental needs, the next time the succession-planning process is reviewed, the individual is still viewed as being two years away from being ready to assume the position. Thus, when specific actions are not taken to meet developmental needs, the succession-planning process becomes little more than a paper exercise that has very limited value.

Another key ingredient of a successful succession-planning system is the verification of a technical manager's assessments of potential successors for various key positions and their readiness to assume such positions. I have witnessed certain organizations that went through a fairly time-consuming, elaborate succession-planning process but did not verify technical managers' assessments regarding potential successors. As a result, the information contained in succession-planning documents, when funneled upward in an organization, was often viewed by top management as having very little credibility.

To avoid this potential problem, it is probably advisable to have each manager in an organization review and "sign off on" the succession-planning information provided by the members of management under her jurisdiction. If a manager and her supervisor do not agree regarding a particular individual's readiness to assume another position in the organization, this matter should be discussed thoroughly, and some type of resolution of this disagreement needs to be made. If this process is followed upward through the organization, the succession-planning information that is distributed to top management should have credibility and be very valuable to the organization.

One final issue concerning succession planning is that of determining whether a technical employee should be shown specific succession-planning charts and other information that contains his name. Some individuals might argue that such information is confidential and should not be shown to individual employees. Others might argue that it is best to be completely open and to show all employees all succession-planning information. I would recommend an approach that is somewhat in between these two extremes. I think it is appropriate to share with employees a manager's assessment regarding some future possibilities with the organization. However, I do not feel that it is necessary, nor in the best interests of the organization, to show employees all succession-planning documents. Instead, I feel that managers of technical employees should have regular career development discussions with their people that provide some general information concerning possible future career opportunities with the organization. Having such discussions demonstrates to employees that their manager is concerned about their welfare and their future. However, because employees are not shown specific succession-planning information, potential conflicts regarding specific employee expectations and what actually happens can be minimized. For example, suppose an employee is listed as being a potential replacement candidate concerning a particular position, but the employee is later thought not to be qualified as a potential replacement. If the employee is not shown the specific succession plan information, disappointment and/or anger concerning the change in the succession-planning document can be avoided.

8.6 TEAM BUILDING

The various developmental activities that have been discussed thus far in this section have focused primarily on the development of individuals. The last developmental activity that will be discussed in this chapter focuses primarily on the development of groups of technical employees. This developmental experience is known as team building.

Team building is one type of activity that is usually categorized under the general heading of organizational development. There are many different team-building approaches, but all share the common objective to improve the effectiveness of groups [5]. These groups could be functional departments (for example, manufacturing engineering), project teams comprised of members from various departments, or different departments that must work together on a regular basis.

A manager of technical employees can do formalized team building himself, or he can involve some other person, either internal or external to the organization, who can serve as a facilitator for team-building sessions [5].

I advocate the use of a facilitator other than the manager of a group. The main advantage of using someone who is not a part of a group as the facilitator for team-building sessions is that such an individual can be viewed by others as independent and unbiased. If a manager of a group of technical employees attempts to formalize team building herself, there is a good chance that at least some members of the group may feel that the manager is not completely unbiased. This perception is likely to occur even if a manager of technical employees is very fair and objective because a manager is a member of the group. As such, others may perceive her as having her own personal agenda, even if the manager is a very fair-minded person.

Whether a technical manager decides to lead team-building efforts herself or to enlist the assistance of another person to serve as a facilitator, I recommend a fairly simple approach to team building. I've used this method on numerous occasions in the past, and it has led to very successful results.

First, I recommend that the facilitator meet with each of the members of the team individually to find out what each person feels is going well and is not working effectively [5]. I tell each person with whom I meet that the content of what is said in the individual meetings will be shared with the group at a later date. However, no names will be associated with the content.

After meeting with all of the individuals who are associated with a particular team-building effort, I summarize all the content of the meetings [5]. On one sheet, I include areas that have been mentioned with regard to things that are going well concerning the group. On a separate sheet, I list the areas that have been mentioned with regard to aspects needing improvement.

For example, in doing team building with a manufacturing engineering department and a product design engineering department, individuals might mention favorable aspects of the two departments' interaction, such as (1) members of both departments are bright; (2) people in of both departments want to do their jobs effectively; and (3) employees in both departments are very knowledgeable regarding their jobs.

Comments concerning areas needing improvement with regard to the interaction of the two departments might include (1) product design engineers do not seem to consider manufacturability when they do their jobs; (2) manufacturing engineers seem to be more concerned about the simplicity of the manufacturing process than about whether a product meets a customer's needs; and (3) members of both departments fail to inform people from outside their department regarding information that is needed.

After the two lists have been made, I meet with all of the people involved as a group. I start the meeting by listing the key points of my first list on a flip chart. These points are then shared, one by one, with the entire group.

Next, the points from the second list are placed on flip chart sheets and shared with the entire group. If any of the group members have questions about any of the concerns listed on the chart, these are addressed.

Group members are then asked to prioritize the list of concerns [5]. People are asked to rank the concerns from most to least important. After a consensus concerning the ranking has been obtained, the highest priority concerns are addressed, one by one.

The group members are asked to brainstorm [6] about possible way of dealing with the top-priority concerns. During the idea generation phase, evaluating the merit of ideas is prohibited. Members are simply asked to "freewheel" and to come up with as many ideas as possible, without stopping to consider the quality or practicality of the ideas [6].

After the idea generation segment is completed for a particular issue, the members are asked to evaluate each of the ideas in terms of, for example, the time involved, cost, practicality, and how well each will solve the problem [5]. The group is then asked to arrive at some sort of consensus concerning the best solution(s) to the problem being discussed. Then one individual is assigned the responsibility of ensuring that the recommended solution(s) is (are) implemented by the dates that are mutually agreed upon by the group [5].

If there is time to address some of the issues in addition to those that have the highest priority, this can also be done at the meeting. Another option is to schedule additional sessions during which other concerns can be addressed, using the process described above.

I have had significant success using the described team-building process. Using it has helped to improve the effectiveness of a number of groups with whom I have worked. Team building is not perfect, nor is it a panacea with regard to all of the problems that exist within a group or between two groups. However, team building can be of significant value in that it can help to improve the effectiveness of groups [5].

8.7 SUMMARY

This chapter discussed some means for developing employees in addition to those described in the previous two chapters. Outside courses can supplement those topics that are addressed by internal training. Although formalized reading programs do not seem to be used in many organizations, they are fairly inexpensive and yet can be quite beneficial. Regular individualized career development discussions do not occur as often as most technical employees would like. When managers hold these discussions regularly, they can have a beneficial impact on both employees and the organization. Succession planning is an essential complement to career planning. Team building focuses primarily on improving the effectiveness of groups of technical employees.

References

[1] Gutteridge, Thomas G., "Organizational Career Development Systems: The State of the Practice," in *Career Development in Organizations*, Douglas T. Hall and Associates, San Francisco: Jossey-Bass, 1986.

[2] Burack, Elmer H., and Nicholas J. Mathys, *Career Management in Organizations: A Practical Human Resource Planning Approach,* Lake Forest, IL: Brace-Park, 1980.

[3] Eppley, W. Robert, and Arthur M. Cohen, *Interactive Career Development: Integrating Employer and Employee Goals,* New York: Praeger, 1984.

[4] Appley, Lawrence A., and Keith L. Irons, *Manager Manpower Planning: A Professional Management System,* New York: AMACOM, 1981.

[5] Dyer, William G., *Team Building: Issues and Alternatives,* Second Edition, Reading, MA: Addison-Wesley, 1987.
[6] Osborn, A. F., *Applied Imagination,* Third Edition, New York: Scribner's, 1963.

Chapter 9

Two Critical Motivators

9.1 INTRODUCTION

As a psychologist, I recognized early in my career that people have significant individual differences. As a result of these differences, what serves as a powerful motivator for one individual may be only moderately or perhaps not at all motivating to someone else.

Despite these individual differences, however, I have come to the conclusion that there are two extremely critical motivators that a manager needs to employ to stimulate others to strive for excellence. I believe that these two motivators are relevant to all employees regardless of their individual differences, and, despite the differences between technical and nontechnical employees, these two motivators are equally relevant to both types of employees. The first essential motivator is setting challenging but realistic expectations. The second critical motivator is demonstrating a true concern for employees.

9.2 SETTING CHALLENGING BUT REALISTIC EXPECTATIONS

The first essential motivator that I believe all managers of technical employees need to use to stimulate their staff members to strive for excellence is setting challenging but realistic expectations. I want to deal initially with the first part, that is, setting challenging expectations. Various authors have indicated that better results are obtained when goals are challenging versus when they are easy to accomplish [1–7].

Managers of technical employees need to set high expectations to achieve outstanding results consistently. If a manager has done a good job of selecting an employee, the individual should tend to set high expectations for himself. Even if an employee has high expectations for himself, however, a technical manager needs to help him "stretch." Many individuals, sometimes even those who have high expectations for themselves, may tend to set constraints that can limit their accomplishments.

Very few, if any, people use their entire potential. Sometimes people talk about "giving 150 percent." Such a statement bothers me, because if 100 percent represents a person's maximum capabilities, she, by definition, can never exceed this figure. However, even attaining something close to 100 percent is something that few people do. It requires pushing oneself to the limits. However, since many people tend to set unnecessary constraints on themselves, they often need someone else who can help them to overcome these constraints. A technical manager can be such an individual.

How does a technical manager help an employee push himself to his limits? I feel that a technical manager can do this by telling an employee that she has confidence that the employee can achieve even greater performance. She might tell her staff member something such as, "This might sound like an unattainable goal at first, but I have confidence that you can do it." Then the manager might help the employee to break down the goal into various subgoals and to determine how each of these could be accomplished.

For example, suppose a manager of manufacturing engineering has an objective for a relatively inexperienced manufacturing engineer that involves changing a particular process because excessive product defects are occurring. Suppose further that the manager asks the engineer to analyze the problem, to develop an improved process, and to train the relevant production employees regarding it within one month.

In this example, the engineer might initially feel that he cannot accomplish the objective. However, the manager could instill confidence by saying first, "I know that you can do this." The manager then might ask the employee a series of questions such as the following.

1. How would you go about analyzing the current production process?
2. How long would it take to do this?
3. How would you develop an improved process that would result in fewer product defects?
4. How long would that take?
5. How would you train the production employees concerning the new process?
6. How long would that take?

If the employee expresses uncertainty concerning any of the questions, the manager could provide her input. At the end of the discussion, the employee should accept that the objective can be accomplished.

When a technical manager is successful in motivating an employee by setting high expectations, the employee often says something such as, "I really didn't think I could accomplish it initially, but my supervisor just insisted that I could do it. I guess she was right."

A manager who is unable to motivate a technical employee toward excellence often agrees with an employee concerning his self-imposed constraints. Some technical managers may indicate disagreement when an employee mentions some self-imposed constraints, but they are unable to convince the employee that he can, in fact, achieve a goal that may initially appear to be unattainable. It is critical that a technical manager truly believe that an employee can accomplish a task. If a manager does not genuinely think that an employee can accomplish an objective, it will be exceedingly difficult, if not impossible, to convince the employee that the goal can be reached.

To illustrate the significant impact that a manager's expectations can have on her staff members' performance, suppose a product design manager for an electronic components manufacturer successfully managed the introduction of two new products in the past year, in addition to managing minor product design modifications in five other products. Suppose further that this product design manager has been doing the job for only about a year and a half and that he is continuing to learn and to improve his job performance. If the technical manager who supervises this individual has only moderate expectations concerning the product design manager's performance, the supervisor might agree with the staff member, if he maintained that he could handle no more than two new product introductions in the forthcoming year.

On the other hand, if the product design manager has a supervisor whose expectations are rather high, his supervisor might disagree that the individual could handle no more than two new product introductions in the forthcoming year. Instead, the product design manager's supervisor might encourage him and indicate that because of the significant amount that he had learned over the previous year, handling three new product introductions in the forthcoming year should be quite attainable.

When I mentioned the importance of setting challenging expectations, I added that these expectations should be realistic. When a technical manager sets expectations that are unrealistically high, the situation rapidly erodes into a very demotivating one [3, 4]. Also as mentioned previously, employees often sell themselves short in terms of what they can accomplish. As a result, an employee may feel that accomplishing a particular goal is not realistic, but a manager may be able to persuade the individual that she does, in fact, have the capability to accomplish the objective.

Thus, I am defining the term "realistic" with respect to what a technical manager genuinely feels is attainable by a staff member. I feel that a true test of what is realistic results when a technical manager asks herself, "If I had to accomplish this objective myself, could I do it?" A second question that a technical manager probably needs to ask herself with regard to determining whether a particular goal is realistic is, "If I were this staff member (that is, if I had his experience and abilities), would I be able to accomplish the objective?"

The following scenario describes a situation that I observed involving a manager who had set unrealistic expectations for one of his staff members. The staff member had not been accomplishing all of the objectives that his supervisor had established for him during a given period of time. Since this situation was going to be causing significant organization problems in the near-term future, the technical employee's supervisor told the employee that he was being relieved of some of his responsibilities.

Shortly after taking over these new responsibilities, the technical employee's supervisor told the employee, "There is no way that these objectives can be accomplished in the time frame that has been established." In this case, it seems clear that the supervisor realized that the objectives were unrealistic, but only after he personally was responsible for accomplishing the objectives within the given time frame. I believe that if the technical employee's supervisor had asked himself the question about being able to accomplish the objectives if he were responsible for doing so himself, he would have amended the expectations for his staff member.

The following scenario illustrates the importance of a technical manager asking the second question referred to previously (that is, "If I were this person, would I be able to accomplish the objectives?"). A new product engineering group for a manufacturing firm is responsible for the introduction of eight new products in a particular year. The director of the group has had extensive experience in new product development. As a result, if she were working as a product development manager, she probably could administer the introduction of four new products within a year in addition to administering minor product changes in existing products.

In assigning the responsibility for new product introduction to her staff members, this individual might be tempted to assign to one of her product design managers the responsibility for introducing of four new products, since this is an objective that she would be able to accomplish, given her level of experience. As long as the individual to whom the responsibility has been assigned is equally experienced in new product introduction, this assignment might be appropriate. However, if a new product design manager has had only minimal previous experience in the introduction of new products, managing the introduction of four new products in addition to the regular work load might be an unrealistic objective.

Thus, it is important for a technical manager to "put himself in an employee's shoes." If a technical manager genuinely feels that he would be able to accomplish a certain objective, given the assumption that he has the same abilities and experience as an employee, then assigning that objective to the employee would probably be appropriate.

To ensure that a technical manager sets realistic expectations, the individual needs to ensure that he does not confuse what he would like to happen with what is realistic. For example, one manager whom I know tends to set objectives based on what he *wants* to occur, regardless of whether this is realistic or not. As a result of this, his objectives are seldom met, and he is frequently disappointed.

Another technical manager I know tends to view objectives that other people might consider to be unrealistic to be a challenge. He seems to employ the philosophy that, "If they say it can't be done, I'll prove that it can." Unfortunately, the objectives that this individual sets for himself and others are frequently not met, and he is often very disappointed.

There are some individuals who believe that expectations or objectives should be set at a level that is so high that they probably will not be met. However, these individuals believe that setting unrealistically high objectives motivates people to put forth considerable effort. As a result, they believe that the objectives accomplished, although they may not reach the level specified, are nonetheless very significant.

I believe that the philosophy/practice described above is inappropriate and feel that it should not be used because it is based on an erroneous assumption: that people will continue to be highly motivated after repeatedly failing to achieve objectives. In my experience, this simply is not true.

I have found that people certainly can learn from their mistakes and failures. In addition, an initial failure can sometimes serve as a motivator to try even harder the next time. However, if a person continually fails to achieve his objectives, this leads to frustration, disappointment, and a lower level of motivation for most people [4]. If people repeatedly fail to accomplish their objectives, they eventually often tell themselves, "These objectives cannot be obtained regardless of how hard I work, so why should I even bother?"

Various research supports my point. A number of studies indicate that goal difficulty and motivation have an inverted-V relationship [2]. In other words, an individual becomes more motivated as the difficulty of goals increases, up to a certain point. However, when an individual feels that goals are too difficult, poor motivation tends to result.

When a technical manager has reasonably high expectations (that is, ones that are challenging yet realistic), there is a fairly high probability that his staff members will meet these expectations. Achieving her objectives motivates a technical employee. It tends to make her feel good about herself, and it gives her confidence that future objectives, even ones that are somewhat more challenging, are also likely to be accomplished. My philosophy, which is consistent with what I have witnessed in my working career, might be simply described by the phrase, "Achieving success leads to future success; repeated failure leads to future failure."

I can provide an example that illustrates that achieving success leads to future success. An engineering manager whom I know was very successful in solving problems for an important customer. He was able to increase his self-confidence and his supervisor's confidence in him because of his success. When problems with other customers arose, he was asked to solve them. He was successful in his efforts to deal with these problems as well.

I can also give an example that illustrates the point made above about repeated failure leading to additional failure. I know a quality engineer who was asked to solve a major quality problem in an unrealistic time frame. He was unsuccessful. As a result, he began to lose confidence in himself and his supervisor lost confidence in his abilities. When he was asked to solve other extremely difficult product quality problems, he was also unsuccessful. This pattern continued for a time; eventually, he was fired from his job.

Some technical managers who believe in setting unrealistically high objectives also believe that if someone reaches an objective, he is then likely to "slack off." In my experience, this simply does not occur. On the contrary, for most people, achieving an objective stimulates continued, and sometimes even increased, effort in the future.

In my experience, I have met a few people who seem to be able to continue to be motivated despite repeatedly failing to achieve unrealistic objectives. However, such individuals are extremely rare. As a result, an approach that may appear to work with them (that is, setting unrealistically high expectations) will simply not work with the majority of technical employees.

For example, a vice president of engineering whom I know used to give unrealistic objectives to his staff members. One individual perceived this as a challenge, and he continued to pursue his goals despite a lack of success. The other staff members, on the other hand, tended to become very discouraged because they continually failed to accomplish the unrealistic objectives. As a result, they appeared to put forth minimal effort. They seemed to feel that there was no point in making exceptional efforts since they were not going to reach the objectives, no matter what they did.

9.3 DEMONSTRATING TRUE CONCERN FOR PEOPLE

In addition to setting challenging but realistic expectations, it is also essential that a technical manager demonstrate a true concern for people to motivate technical employees toward excellence [3, 8, 9]. I have known many managers who are skillful at motivating others, and their management styles vary greatly. Many of these managers use a very participative approach, emphasizing considerable employee involvement. Others are very hard-nosed, tough-minded, somewhat autocratic managers. A common characteristic shared by all of these excellent motivators, however, is that employees believe that they truly care about their staff members' welfare. I have observed many technical employees who were willing to work extremely long hours and make many personal sacrifices because they felt that their supervisor really cared about them as people. I believe that technical employees tend to view the employment situation in the context of a social contract [9, 10].

As such, if an employee feels that she is valued highly by her supervisor as a person, then the employee feels that she owes something to her supervisor in return: to do everything within her power to accomplish the job-related objectives.

On the other hand, a technical employee reacts quite differently when the individual feels that his supervisor does not particularly care about him as a person. In such a situation, a technical employee is likely to feel no significant obligation to "perform above and beyond the call of duty."

I believe that when a technical manager who does not appear to care a great deal about his employees has very high expectations concerning employees' performance, the employees tend to view this as an inequitable situation. As a result of this perceived inequity, a technical employee in this situation is likely to take one of two actions; that is, she may either resist putting in a great deal of extra effort on the job, or the individual may begin looking for an alternative employment situation that is more equitable from her viewpoint [6].

The following example involving an engineer who worked for a manufacturing firm illustrates the points that I made. This individual had an extremely demanding job involving exceedingly high and probably unrealistic expectations on the part of his supervisor. The engineer was under a great deal of stress and felt that his relationship with his family was being adversely affected as a result of the extremely long hours that he was working. He finally decided to resign from his position. However, he said that he probably would have stayed, had he felt that his supervisor truly appreciated his efforts and cared about him as a person.

Another example serves to illustrate a quite different situation. In this case, another engineer had an extremely demanding job and was working exceedingly long hours. This individual also felt a great deal of stress, and significant job demands were creating pressure with regard to his family relationships. Despite the situation, however, this person continued to stay in his job. He said that he remained in his position because he knew that his supervisor appreciated his efforts and cared about him as an individual.

What are some examples of actions that a manager of technical employees can take to demonstrate his genuine concern for the employees' welfare? One very effective technique is to make special efforts when an employee needs help. For example, the vice president of technology for a manufacturer of equipment came in on a Sunday evening to help some engineers who were working hard to meet a new product introduction deadline. One of the engineers told me that this action meant a lot to him because it demonstrated that the manager truly cared about his welfare.

Another effective action a manager can take is to ask an employee what she can do to make the individual's job easier. A manager of research for one company asked this of his employees on a regular basis. Several of them told me that this was greatly appreciated.

Demonstrating support is an additional important action that a manager can and should take. For example, the manager of process engineering for one company intervened on behalf of one of his employees when he heard that a manager of another department was complaining about the employee. He went to the other manager and asked about his concerns. He then explained why his staff member had done what he had, and he indicated that he supported the actions.

Another example of demonstrating support is making sure that technical staff members have the resources that they need to do their jobs. For example, the vice president of engineering for one manufacturing company submitted a budget for new hardware and software for several of his people. When the president of the company cut these items from the budget, the vice president met with him and defended the need for the new software and hardware. He persuaded the president to leave these items in the budget for the upcoming year.

Asking about an employee's relative who is ill is an additional example of an action that can demonstrate concern for employees' welfare. Keeping track of the names of an employee's spouse and children and their interests and activities, and commenting on them from time to time also demonstrates a true concern for an employee.

The president of a fairly large company used to keep a notebook containing relevant information about his key employees' families. He freely admitted that it was very difficult to remember such information without the memory aid. Several employees mentioned that they felt very good when the president asked them about their spouse and children, mentioning their names and some of their activities.

Suggesting that an employee take some time off, particularly after she has completed a major project involving substantial extra hours, is another way to demonstrate concern for an employee's welfare. One project manager for an engineering consulting firm did this after one of his staff members completed a major project that he had worked on seven days a week for several months. The employee took a vacation with his spouse, and the spouse sent the manager a thank you note for suggesting that her husband take some time off from work.

What is not effective in demonstrating a genuine concern for employees' welfare? One good example is simply paying lip service to the concern for staff members. "Actions speak louder than words" is a good phrase to remember in this regard. A manager's actions must reinforce his words.

For example, I know one manager who used to tell his employees on occasion that he was concerned about their welfare. One evening, one of his technical employees had to work until 3 AM to complete a project by the deadline. The manager had reminded the individual that the project was due the next day. However, when he had left at 6 PM, he had not offered to assist the employee in any way. After this occurrence, the employee doubted whether the manager was really concerned about his welfare.

9.4 SUMMARY

Despite numerous individual differences among technical employees, they also share some commonalties. By taking into account the similarities among their staff members, technical managers can identify motivators that are common to technical employees in general. The two most critical motivators, in my opinion, are (1) setting challenging but realistic expectations, and (2) demonstrating true concern for people. I feel that if a technical manager fails to use either of these two motivators with her people, she will not be able to stimulate excellent performance from her staff members over time. If, on the other hand, a manager of technical employees uses both of these motivators effectively, she will be able to make significant progress in stimulating top-level performance from her staff members. Additional important, but perhaps not as critical, motivators will be discussed in the next chapter.

References

[1] Champagne, Paul J., R. Bruce McAfee, *Motivating Strategies for Performance and Productivity: A Guide to Human Resource Development*, New York: Quorum, 1989.

[2] Chung, Kae H., *Motivational Theories and Practices*, Columbus, OH: Grid, 1977.

[3] Dessler, Gary, *Improving Productivity at Work: Motivating Today's Employees*, Reston, VA: Reston, 1983.

[4] Green, Thad B., *Performance and Motivation Strategies for Today's Workforce: A Guide to Expectancy Theory Applications*, Westport, CT: Quorum, 1992.

[5] Locke, Edwin, and Gary Latham, *Goal Setting: A Motivational Technique That Works!* Englewood Cliffs, NJ: Prentice-Hall, 1984.

[6] Saal, Frank E., and Patrick A. Knight, *Industrial/Organizational Psychology: Science and Practice*, Pacific Grove, CA: Brooks/Cole, 1988.

[7] Steers, Richard M., and Lyman W. Porter, *Motivation and Work Behavior*, Second Edition, New York: McGraw-Hill, 1979.

[8] Dessler, Gary, *Winning Commitment: How to Build and Keep a Competitive Workforce*, New York: McGraw-Hill, 1993.

[9] Kushel, Gerald, *Reaching the Peak Performance Zone: How to Motivate Yourself and Others to Excel*, New York: AMACOM, 1994.

[10] Thomas, R. Roosevelt, "Harvard Business School Note: Managing the Psychological Contract," *Manage People, Not Personnel*, Harvard Business School, Cambridge, MA, 1990.

Chapter 10

Other Key Motivators

10.1 INTRODUCTION

The two critical motivators discussed in the previous chapter must be used by a technical manager if she expects to maintain excellent performance over time. However, there are also other important motivators that a technical manager needs to consider. Although they may not be quite as important as the two motivators discussed in Chapter 9, they are nonetheless very significant.

10.2 PROVIDING INTERESTING, INTELLECTUALLY STIMULATING WORK

Most employees enjoy work that is interesting and intellectually challenging [1]. This is particularly true about technical employees. I believe that one of the major reasons why many people decide to pursue a technical career is that they find that technical work provides them with frequent opportunities to challenge their intellect.

As a result of technical employees' need for frequent intellectual challenge, a technical manager needs to ensure that his staff members have the opportunity to experience intellectual stimulation in many of the tasks that are assigned to them.

How does a technical manager ensure that the majority of the tasks assigned to an individual are intellectually challenging? Probably the best way is simply to ask the employee. Every job has aspects that are more mundane and less stimulating than other job tasks. However, employees should feel that at least the majority of the tasks that they do are intellectually satisfying.

A technical manager needs to recognize that because of individual differences, what might be intellectually stimulating to one person might be rather boring to someone else. For example, building a prototype of a new product might be somewhat boring to an engineer, particularly one who has had considerable experience in the past regarding the building of prototypes. On the other hand, this task might be very challenging to a laboratory technician, particularly one who has had only limited exposure to this in the past. Technicians tend to be motivated by building prototypes because they typically like to work with their hands and they often enjoy making things work.

Another key point that a manager of technical employees needs to keep in mind is that what might be intellectually stimulating today could become rather mundane and boring tomorrow. As a result, it is important for a technical manager to have repeated conversations with her employees concerning the level of intellectual stimulation that they find in their job tasks.

For many employees, after they have done a particular task numerous times, they may consider the task to be rather routine and perhaps even somewhat boring. For example, an equipment designer, after having designed over one hundred machine guards for various pieces of equipment, might very well find that designing another machine guard for a new piece of equipment would be a rather mundane and not particularly challenging task. This might be the case, despite the fact that the same individual may have found it to be very interesting to design the first few machine guards for various pieces of equipment. Interest in this task might be maintained or enhanced by letting the equipment designer develop a CAD/CAM application that does the detailed design of a machine guard. This would allow the company to avoid losing his specific experience base.

The time interval between repetition of a particular task can also be important. For example, a consulting engineer might be asked to provide a cost estimate for the construction of a manufacturing facility. If this individual were asked to develop cost estimates for other manufacturing facilities every week for the next twelve weeks, she might find the task to be more mundane each time she performs it. One way to deal with this might be to automate the task using a computer application. Alternatively, doing the cost estimates could be delegated to another employee who has not done this before and wants to learn something new.

One final point is that a technical manager needs to balance providing challenging tasks with getting the work done by the most capable individual. It is important for a manager to take advantage of his staff members' expertise to achieve maximal results. Occasionally doing this may conflict with giving employees a chance to learn new tasks and be intellectually challenged. For the most part, however, a technical manager should be able to do both (that is, to take advantage of his staff members' greatest talents and to provide them with opportunities to be challenged intellectually).

10.3 GIVING APPROPRIATE FEEDBACK

Another key motivator for technical employees is providing them with appropriate feedback concerning their performance [1, 2]. People want to know how they are doing. One of the significant causes of stress and anxiety is uncertainty. If a technical employee does not know how her supervisor is evaluating her job performance, she may spend a great deal of time pondering, and perhaps even worrying about, her supervisor's evaluation. On the other hand, if a technical employee's supervisor gives her feedback about her performance, there is no uncertainty, and, as a result, the employee's stress and anxiety can be greatly minimized. After receiving specific feedback, the employee can continue her performance with regard to tasks that her supervisor feels are being done properly. In addition, the employee can make changes in performing tasks that her supervisor indicated need improvement.

Employees vary regarding the amount of feedback they desire or need. Individuals that have a strong need for feedback, new employees, and people who are having performance problems need to receive feedback considerably more often than others. Thus, if a manager of technical employees is going to err, it is best to provide more feedback than is needed rather than less. Unfortunately, most technical managers probably tend to err in the direction of giving feedback too infrequently.

When certain technical employees ask for more feedback, technical managers are sometimes a bit confused. For example, I have heard comments such as the following from technical managers who have been asked to provide additional feedback to employees: "I don't know what more she wants. We just went through a performance review three months ago."

Unfortunately, conducting a formal performance review with an employee does not obviate the need for ongoing, less-formal feedback. In addition, even though a manager provides ongoing, informal feedback to a technical employee, the manager still needs to provide formal performance reviews at regular intervals [1].

Ongoing informal feedback and formal performance reviews serve different purposes. Informal feedback can help to keep an employee on track with regard to his performance on an ongoing basis. Formal performance reviews should merely be summaries of informal feedback that has been given over the past appraisal period [1]. This formal, summary data serves to reinforce the ongoing messages provided by technical managers. Without regular, formal performance reviews, technical employees sometimes do not fully recognize the significance of their supervisor's feedback. On the other hand, without ongoing, informal feedback, a technical employee sometimes forgets the summary information provided during formal performance reviews.

When giving a technical employee feedback, whether this is a part of a formal performance review or an informal discussion, the manager needs to be as specific as possible [2]. For example, if a manager says, "You need to improve your communication," the employee is unlikely to understand what this means. It could mean, for example, that the employee does not send copies of her memos to the technical manager. The statement might also mean that the employee does not write clearly. Alternatively, the statement might mean that the technical employee is too verbose in her discussions.

A much more appropriate and considerably more specific statement that the technical manager might make to his employee is, "When we have meetings with the marketing department concerning the status of products currently being developed, you tend to provide much more detail than is needed. For example, at the last meeting, Bill asked when the new pressure gauge would be ready and you, consequently, spent fifteen minutes explaining the latest technical difficulties of the product. All he wanted was an estimated date of when the prototype was expected to be ready."

The following scenario provides another example of how to make performance feedback more specific. Suppose a manager tells an employee, "Your attitude needs to improve." This might mean that the employee does not cooperate with coworkers. It might also mean that the quality of the employee's work needs improvement. The statement could have other meanings as well; it is simply much too vague.

What could the manager say that would be more specific and more meaningful? One example might be, "When coworkers have asked you for assistance in the past few months, you have often indicated an unwillingness to help them. For example, last week, Bill asked you for help regarding the new product launch. He said that you told him you were too busy with other work."

10.4 REWARDING TOP PERFORMANCE

One of the most powerful and most underutilized motivators of technical employees is providing positive feedback concerning excellent performance [1, 3]. When a technical employee receives some form of positive feedback from her supervisor regarding outstanding job performance, this makes her feel good about herself. In addition, when a technical employee receives positive feedback, she typically feels that her efforts were worthwhile, and she is also likely to spend significant effort in the future to get additional positive feedback.

Positive feedback can take many forms. On the one hand, it may consist of a simple comment such as, "Mary, you did a great job on that presentation yesterday." At the other end of the spectrum, a technical employee who did an outstanding job on some significant project might be given a special award and an engraved plaque at a formal company-sponsored dinner.

Most positive feedback is fairly simple [1] and does not require much time to provide. Typically, the cost is negligible [1] and, as I mentioned previously, it can have a dramatic impact on employees.

Despite this, however, providing positive feedback is a very underutilized motivator in most organizations [1]. In fact, although I have worked with many different organizations throughout my career, I would have a difficult time identifying more than one or two that I would rate highly in terms of providing positive feedback on a frequent, ongoing basis to employees who demonstrate outstanding performance.

One organization for which I worked early in my career was the Parker Pen Company. At least once every month or two when I worked at Parker, I received some sort of spontaneous, informal compliment about my work performance. For example, my immediate supervisor mentioned that my abilities in the area of selecting key employees were excellent. One of the heads of a major division told me that my individualized developmental counseling of key employees was extremely helpful. A number of attendees at my management development seminars indicated that the sessions were very beneficial to them. Receiving positive feedback on a fairly regular basis from various individuals in the organization was a powerful motivator for me. Such comments made me feel greatly appreciated and important to the organization. They also added to my enjoyment of my job immeasurably. In addition, they stimulated me to continue to strive for performance excellence.

Spontaneous comments such as these have a significant effect on virtually everyone. B. F. Skinner, a well-known psychologist, recognized the powerful impact of compliments and other forms of positive reinforcement. He noted that if an individual receives some type of positive reinforcement after she has exhibited effective performance, the likelihood of similar performance occurring in the future is increased [3]. Organizations that recognize and use Skinner's principles regularly are capable of fostering superior performance on an ongoing basis.

Why don't most technical managers use positive feedback as a motivator more often? I believe that there are a number of major reasons, which include the following:

1. Technical managers are often preoccupied with their work and just do not take the time to think about providing positive feedback to employees who have demonstrated excellent job performance [1].
2. Technical managers may simply not feel comfortable about providing positive feedback to one of their staff members because they are not as communicative as others in the business world.
3. Most technical managers do not receive a great deal of positive feedback from their supervisors and, as a result, do not have a good "model" to imitate.

4. Some technical managers may have good intentions, in that they plan to provide some positive feedback to an employee who has done a good job, but then may forget to do so when some other, more urgent matter arises.
5. Some technical managers may even feel that providing positive feedback is not necessary or appropriate [1]. For example, when I told a group of technical managers at a meeting that I felt that they needed to provide more positive feedback to employees, I was met with a number of blank stares. Then, one technical manager asked, "Do you mean to say that you think that we can improve the performance of our people by giving more positive feedback?" When I answered this question affirmatively, the nonverbal response that I received from the group indicated that they felt that I had just provided a rather bizarre suggestion. At the time, I almost felt that I would have gotten a similar response had I suggested that they should all sell their children and use the money to go on a great vacation.

There may be additional reasons why technical managers provide infrequent positive feedback to their staff members. The key point is that, regardless of the reasons, the benefits gained from the use of more frequent positive feedback greatly outweigh the costs.

In addition to providing positive feedback to employees who do an outstanding job, technical managers also need to provide generous financial rewards to such individuals [3–5]. These rewards could be in the form of significant merit increases and/or substantial bonuses of cash or company stock.

Chapter 1 discussed the importance of providing excellent compensation to attract top-flight employees to an organization. Without repeating the rationale described in Chapter 1, I would simply say that providing excellent compensation can be a significant motivator to current as well as potential employees. For example, one engineering manager whom I met told me that he had received a cash bonus equal to about fifty percent of his base salary because the members of top management in his organization felt that he had achieved outstanding results. This bonus was particularly noteworthy because the total bonus pool paid out during that year was substantially less than usual due to dismal overall corporate financial results.

Although this particular engineering manager was not someone who appeared to be particularly motivated by money, it was clear that he felt that the significant bonus he received represented a very tangible indication to him that his organization greatly valued his contributions.

A technical manager who is very conservative, and perhaps even miserly, in terms of the financial rewards that he doles out to outstanding performers is very foolish, in my opinion. He may save a few dollars for the corporation, but the manager is taking a great risk if an outstanding performer decides to leave the organization. The cost of this loss of a top performer will greatly outweigh any minor benefits obtained by having paid this person less than she might have been compensated.

Some technical managers may try to rationalize the situation when an underpaid outstanding performer leaves the organization by saying something such as, "People do not decide to remain with or leave an organization just because of money. Therefore, even if I had been paying him more, the employee still would have left the organization."

I agree with the first part of the manager's statement that people typically do have multiple reasons for deciding to remain with or leave an organization; however, I do not necessarily agree with the second part of the manager's statement (that is, the employee would have left even if he had been paid more money). Money is a rather complex motivator. It not only can be used to buy items; in addition, for many technical employees it symbolizes the value they feel the organization places on them. As a result, a technical employee's compensation level can easily be the deciding factor influencing a technical employee who is in the process of deciding whether to remain with or leave his current organization.

"Cognitive dissonance" is a term used in the field of psychology. The accompanying theory that goes with this term was initially developed by an individual named Leon Festinger [6]. According to cognitive dissonance theory, if an individual feels that there is adequate justification for one of her particular actions, then the person does not feel any dissonance (that is, discomfort). However, when there does not appear to be adequate justification for an action, the individual does feel dissonance. When she feels this cognitive dissonance, the person is motivated to either change her attitude concerning the action or to take a different action.

Cognitive dissonance theory can be applied to the situation involving a technical employee who is an outstanding performer. Such an individual inevitably expends considerable effort and works very long hours. Often such individuals have at some point asked themselves a variation of the question, "Why am I working so hard?" A highly paid top performer might answer the question with, "I am being paid extremely well, so I guess I am being rewarded and recognized for my efforts." Such an individual would, according to cognitive dissonance theory, feel no particular dissonance, and as a result, the individual would not be likely to leave her current organization.

On the other hand, an outstanding performer who is not particularly well paid might answer the above question in a different way. Such a person might tell himself something such as, "I don't really know why I am working so hard. I'm not paid particularly well, and I don't feel that the organization recognizes my value." According to cognitive dissonance theory, such an individual is likely to feel substantial dissonance. To reduce this dissonance, the person can either try to change his attitude concerning his behavior or he can change his behavior. Trying to convince himself that he really is not working especially hard would probably be rather difficult to do. Deciding to work less hard would be a very difficult decision for a typical outstanding performer to make. The other alternative is for the technical employee to decide to look for another position in an organization that is likely to reward and to recognize him for the significant effort expended and results obtained.

10.5 ELIMINATING POOR PERFORMANCE

If a technical manager does an excellent job of selecting employees of the highest caliber, she will not need to devote much time and effort to dealing with poor performers. Ideally, a technical manager should not have to deal with this issue at all. However, realistically, most technical managers need to deal with occasional poor performers, even if they are extremely effective in selecting employees.

First of all, unless a technical manager is starting a new department, she is likely to "inherit" employees who were formerly supervised by someone else. Unfortunately, in many such situations, a technical manager will inherit at least one employee whose performance needs considerable improvement.

In addition, even a technical manager who has excellent selection abilities is likely to make an occasional mistake in hiring. Also, some employees who are superior performers may go through some difficult periods in their lives during which their performance on the job drops considerably.

Thus, most technical managers will need to deal with the issue of poor staff member job performance at least occasionally. Unfortunately, this is an area that many technical managers do not like and are not particularly adept at handling.

Many technical managers, when faced with a performance problem regarding a staff member, initially do nothing. They may sometimes tell themselves, "I hope that the employee can resolve this problem on her own." Unfortunately, this is merely wishful thinking about 99 percent of the time.

It is essential for a technical manager to point out a significant performance problem immediately. The manager, first of all, owes it to the employee to be candid regarding his performance [1]. This gives the employee an immediate opportunity to try to do something about the performance problem. Although the manager's comments in this regard may not come as a surprise to many technical employees, some individuals may not be aware of the significant performance deficiency. Obviously, for someone to correct a problem, one must first be aware that a problem exists.

A manager of technical employees also owes it to the other members of her team to point out any performance deficiency immediately. People who work very hard and achieve excellent results typically do not appreciate their supervisor overlooking a significant performance deficiency in the unit. If employees feel that their supervisor is not dealing with a performance problem, they may conclude that performance really does not matter. As a result, they may decide that there is no reason to put forth the extra effort that is necessary to attain outstanding results.

Alternatively, some employees who have extremely high performance standards ingrained in them may not feel comfortable in relaxing their own performance expectations, even if they feel that their supervisor's lack of action regarding a performance problem suggests that performance really does not matter. Such individuals probably would continue to demonstrate excellent performance, but they would be likely to feel that the situation is unfair. As a result, they might begin looking for another position in an organization that expects high-level performance and does not tolerate poor performance.

If a technical manager decides to deal with a performance problem, what is the best way for him to address the issue? First, it is essential for a technical manager to explain in detail why an individual's performance is not meeting his expectations [1, 7]. It is important to deal with specific behavior rather than to provide only generalities. For example, if a product design engineer said, "You need to be more careful" to a lab technician who worked for him, it would not be particularly helpful. An example of a much better statement might be, "When you tested the last prototype, you did not record some of the results of the tests." The latter statement tells the employee exactly how her performance fell short of expectations. If the product design engineer added, "We must record all of the test results," this clearly indicates to the employee how her performance needs to be changed in the future.

After a technical manager has clearly explained why an employee's performance is unacceptable and specifically how an employee's behavior needs to be changed, it is important to let the employee know that he is willing to provide any assistance that is needed to correct the performance deficiency. By indicating this, it is likely that the employee will feel that her supervisor is taking a very supportive stance regarding her performance deficiency.

It is also important for a technical manager to indicate clearly a specific time frame during which the manager expects the employee to improve her performance [1]. A manager might, for example, tell an employee that he expects to see a significant improvement with regard to the performance deficiency within the next thirty days. The manager might add that he will meet with the employee after a month to review the individual's progress at that point and to discuss any appropriate future action [1].

As a followup to the meeting, a technical manager might confirm in writing what was discussed. This action tends to reinforce the content of the meeting. It also helps to ensure that the employee understands and is in agreement with the action expected by the manager and the timetable that will be used. Obviously, if the employee does not understand or agree with any of the points mentioned in the written communication, the manager and employee need to meet again as soon as possible to resolve these misunderstandings or disagreements.

After a manager and technical employee have an agreement concerning the action that will be taken to address the performance deficiency, the manager needs to ensure that he does, in fact, have an opportunity to observe the individual's performance and to provide any required assistance. Prior to the next meeting, the manager needs to evaluate the employee's performance. At the meeting, the manager should give the employee some feedback concerning her evaluation.

If an employee's performance dramatically improves and is now totally acceptable to her manager, nothing more formally needs to be done. However, the manager probably should informally continue to monitor the employee's performance, just to be sure that it does not erode to the former unacceptable level. If, on the other hand, an employee's performance improves according to the plan but is not yet at the final expected level, additional sessions need to be scheduled. It is important for the manager to indicate exactly when such sessions will be held and what continued performance level increases will be expected. As long as the employee's performance continues to improve according to the plan and ultimately reaches the final expected level, the manager does not need to take any additional action.

However, if a technical employee's performance does not continue to improve according to the plan established, a technical manager needs to decide what action will be taken and to communicate this to the employee.

For example, suppose a technician is responsible for repairing production equipment that is not functioning properly. Suppose also that this technician frequently experiences problems with equipment that he has repaired (that is, it often still does not work properly after he has supposedly repaired it). This individual's manager might sit down with him initially and indicate precisely in what ways his performance is failing to meet expectations. In addition, the manager might indicate the specific objectives that he is expected to reach in the future.

For example, this individual's manager might tell him:

1. It is unacceptable that the equipment that he supposedly repaired fails to operate properly.
2. He expects that the equipment repaired by the technician will operate properly after the technician's work is completed.
3. He will be meeting with the technician weekly for the next 60 days to review his job performance.
4. To remain in his position beyond 60 days, the technician must ensure that all of the equipment that he repairs works properly after he has completed his work on it.
5. His current performance is expected to continue to improve each week until he reaches the ultimate goal of no equipment failure. In other words, if, for example, ten percent of the equipment on which he works now fails to operate properly after he completes his repair, this percentage is expected to go down weekly until it reaches zero percent after sixty days (or earlier).

During this period of time, the manager must ensure that he is providing the technician with all of the assistance that he needs to improve his performance. This may involve, for example, extensive one-on-one coaching and training.

At the end of the 60-day period, the employee will be well aware of the fact that he either has or has not met his supervisor's performance expectations. If, in this case, the technician fails to reach the performance objectives set for him, the manager needs to discuss other options. For example, the employee might be willing to accept another position internally that is less demanding and more consistent with his abilities. If the manager agrees that the employee could perform such a position successfully and a relevant opening exists, the employee could be transferred. On the other hand, if the employee is not interested in taking another position within the organization, if the supervisor does not feel that he is qualified for another internal job, or if no such opportunity exists within the organization, the manager will need to discuss the subject of termination with the employee.

Although an employee who is terminated is typically rather disappointed, she should not be surprised if the procedure outlined above has been followed. Most terminated employees should not be angry either, as long as they feel that the performance expectations were reasonable and that their manager made every possible effort to help them to succeed. If a technical manager also offers his genuine assistance to an employee regarding her attempts to find another position, the chances that a terminated employee will decide to take legal action against the manager and/or the company may be lessened. Obviously, even if the procedures outlined above are followed, the risks associated with some type of legal action cannot be eliminated entirely. However, a manager who has followed the described procedures should feel that he has taken appropriate action under the circumstances.

10.6 USING PEOPLE'S TALENTS

"Employee involvement" has become a very popular term in recent years. This concept is very appropriate, but it is hardly new. Back in the middle to late 1970s, I designed a management development curriculum for the Parker Pen Company. At that time I "preached the gospel" of what was then called "participative management." Although I believe that many of the managers who attended my seminars at the Parker Pen Company saw some merit in participative management, few people became true believers in this philosophy.

I believe that the recent popularity of employee involvement is due to the many "success stories" that have been reported in which the concept has played a key role. Many managers today finally have embraced the concept of employee participation/involvement because they realize that it works. Employee involvement is effective because it is based on a variation of the old saying that "two heads are better than one." The new version of this proverb might be that "many heads are better than one."

No manager, regardless of how talented he is, knows everything or always has the best ideas. As a result, virtually every manager can benefit from getting input from his staff members. In this regard, it is important for a manager to listen attentively when a staff member provides an unsolicited idea. In addition, however, every manager needs to solicit input from his employees and to use the ideas as much as possible [4]. By doing so, a manager can take advantage of all of the talent that exists in his group.

Thus, probably the most significant and pragmatic reason for the act of solicitation of employees' ideas is that doing so results in better overall ideas and, as a result, improved group performance [8]. In addition, however, there are other important reasons why a manager should actively request ideas from staff members. First, when a manager asks for a staff member's ideas, the employee tends to feel important. In essence, the employee is likely to tell herself that the manager places so much value upon the employee's input that he specifically asked for it.

Second, when an employee is regularly asked for her input, the individual tends to feel that she has significant influence upon the group [8]. Thus, instead of feeling relatively powerless and unable to control her fate, an employee tends to feel that she possesses significant power and control.

Third, by providing regular input to her supervisor, an employee tends to take considerable "ownership" in the group [8]. As a result, the individual often tends to make references to "my group" or "our group" rather than "Jill's group" or "Bill's group." In addition, the employee typically is much more concerned about the quality of her follow-through and that of the other group members regarding tasks.

Regarding the solicitation of staff members' ideas, a technical manager can choose to do this formally or informally. For example, when a technical manager is talking to an employee about a particular problem that the unit is facing, the manager can informally ask if he has any suggestions for dealing with this issue. Alternatively, a technical manager may employ a more formal process for soliciting employees' ideas. For example, at each staff meeting, a technical manager might allot a specified amount of time for brainstorming [9]. This process was described in Chapter 8.

Regardless of whether a technical manager solicits ideas from staff members informally or formally, it is important that she continuously solicit input. If a manager simply mentions that she is interested in receiving staff members' suggestions one time, many employees may not share their ideas because they may not be certain whether the manager is truly interested in their suggestions. Frequent solicitations concerning staff members' ideas obviously takes time, but it is time well spent.

Some technical managers may be reluctant to solicit input from staff members because they view solving the unit's problems as their responsibility. It is true that when a technical manager thinks of a way to solve one of the problems in the unit herself, the manager gets a feeling of accomplishment. However, a manager who is able to solve a problem as a result of soliciting a solution from one of the unit's members can also experience a significant feeling of accomplishment. By asking for input from staff members, a technical manager is using various talents available to her as opposed to using only her own personal resources. A manager can and should get as much, and perhaps even more, satisfaction from solving a problem by using the entire group's talents than from solving a problem by herself. Becoming comfortable in soliciting input from staff members simply requires that a technical manager redefine her role from that of an individual contributor to that of a coach.

There may be some organizations that expect the manager to be the primary person who comes up with new ideas. Even in these organizations, however, results speak for themselves. In other words, if a technical manager demonstrates that he can get even better results by soliciting input from staff members, top management is likely not to complain about his methods.

One final point is that encouraging input from staff members does not mean that every decision will be agreed upon by everyone. There may be times when a staff member, or even several employees, may recommend a particular action that the manager may need to veto. A technical manager, as the head of the unit, has the final approval on decisions and actions.

Ideally, consensus will be achieved on the majority of key issues. However, this will not always be possible. In addition, there may be some cases in which a manager does not solicit ideas from her staff members. For example, in some situations, there may not be sufficient time to get input from everyone. In other cases, an issue may not be significant enough to justify discussing it with all staff members.

10.7 BUILDING A TRUE "TEAM"

Another key way that a technical manager can motivate his employees is to build true "team spirit." The word "team" is a greatly overused term today. Nonetheless, if a technical manager wants to obtain and to sustain excellent performance in his group, it is critical for him to be able to build a true team successfully.

When an outstanding performer is a member of a solid, cohesive unit, this individual is, in a sense, doubly motivated to achieve excellent results. First, the individual is motivated to achieve outstanding results because of individual factors (for example, recognition and a sense of personal accomplishment). In addition, however, an individual who belongs to a close-knit group is motivated to achieve excellence for the good of the group as a whole [10, 11].

Being motivated to perform extremely well for the good of the group as well as for oneself is important for a number of reasons. First, when a person faces a particularly trying situation, individual motivation may sometimes not be enough to sustain personal efforts. In such cases, by reminding oneself of the importance of the group's objectives, a person may be able to continue to sustain substantial effort despite experiencing very trying circumstances.

Another reason why it is important for a manager to build a strong, cohesive team is that doing so may stimulate employees to set objectives even higher than they would solely as a result of their own personal motivation. A cohesive group can, in a sense, build upon one another's successes. In other words, when one member of the group is successful in accomplishing a challenging objective, this may stimulate others in the group to set equally demanding goals. In some cases, these goals may be higher than those that they may have set on their own, if they were not part of the cohesive group.

How does a technical manager build a cohesive team to gain employees' complete cooperation, loyalty, and dedication to the achievement of their individual as well as the group's objectives? One fairly simple yet effective approach that I recommend was described in Chapter 8 and, hence, will not be repeated here.

10.8 LISTENING AND DEALING WITH PROBLEMS

Another extremely powerful motivator that a manager can use with technical employees is simply to listen [1, 7]. Some readers may be somewhat surprised upon reading the previous point because they may feel that listening is a very basic activity that all managers do on a daily basis. However, although it is true that every manager needs to listen each day, very few do an effective job in this area. As a result, a technical manager who is an extremely effective listener stands out among his peers and is able to motivate employees well.

Technical managers usually hear what staff members tell them, but hearing is not the same as listening. Hearing is merely a passive process. It involves the receptance of sound signals in the brain. Listening, on the other hand, is an active process. Effective listening requires attention and concentration.

Why don't technical managers listen more effectively? There are probably a number of different reasons. First, there is a mismatch between peoples' rate of speech and their rate of comprehension. The average individual can listen to speech at a much faster rate than the speed at which people talk. As a result, when an employee is talking to a technical manager, he can listen and still have extra time for additional mental processes. What do most people do with this additional time? Unfortunately, they tend to use it for activities that detract from the listening process. For example, many people tend to think about another subject or about what they plan to say in response to what is currently being said to them.

Thinking about another subject concerns another common reason why technical managers do not listen well: they are distracted. A typical technical manager is very busy and has many subjects on her mind. As a result, it is very easy to start thinking about other important subjects when one is being discussed [1]. For example, a product design engineer might be talking to his supervisor about a problem that he is having developing a new product. Although his supervisor may be interested in this problem, she may have another meeting in an hour with her supervisor concerning a different new product. As a result, her attention may tend to wander from the subject at hand.

Another reason why many technical managers may not be good listeners is that they tend to formulate their response to what is being said while another person is talking [1]. This is particularly true when the manager disagrees with the point being made. For example, suppose a manufacturing engineer is talking to his supervisor about a piece of equipment that he feels should be ordered. In addition, suppose that the engineer's supervisor, the manufacturing engineering manager, does not agree that this particular piece of equipment needs to be ordered. In such a case, the manufacturing engineering manager, while listening to the engineer's argument, may be preparing his counterargument as to why the equipment is not needed. By doing this, the manufacturing engineering manager is probably not listening carefully to the points being made by the manufacturing engineer.

Another reason why many technical managers may not be very good listeners is that they do not know how to listen. Listening is a very important activity that all of us must do every day. Despite this, however, very few people have ever had any formalized training in how to listen effectively. As a result of this lack of training, many individuals, particularly those whose natural listening ability may not be particularly great, do not listen very well.

Here are some pointers that can help a technical manager to improve his listening ability. First, a technical manager needs to give his full attention to the task of listening [7]. A technical manager needs to concentrate completely on listening to what technical employees have to say. This means that when a technical employee is talking, a manager must use all of the time available to focus on what is being said rather than attempting to listen and to engage in some other cognitive processes at the same time.

Although a technical manager is probably always going to have more than one subject on her mind at a given time, there are probably times when a manager is particularly distracted. For example, a half an hour prior to a very important meeting with a technical manager's supervisor might be best spent preparing for the meeting. If one of the manager's employees wants to talk to the manager just prior to such a meeting, the manager might suggest that the discussion be postponed until later so that she can give full attention to what the employee has to say.

A manager must make a special effort to combat against a natural tendency to begin to formulate a counterargument when presented with an argument with which she disagrees [1, 7]. Instead of preparing a counterargument, a manager needs to ensure that she fully understands an employee's position. One way to do this is to ask a number of questions so that a manager has a complete understanding of the various reasons why a technical employee has taken a particular position.

One technique that can help to improve a technical manager's listening ability was referred to in Chapter 3. This approach involves paraphrasing the content of someone's message. When a technical manager is able to paraphrase what an employee has said accurately, this demonstrates that the manager has been listening carefully [7]. The employee in this case usually feels very good because the manager has demonstrated that the employee's message was important enough to warrant the manager's full attention. Even if a manager occasionally incorrectly paraphrases what an employee says, this is not necessarily bad because the misperception can be easily corrected.

Another approach that can aid a technical manager in improving his listening ability is to take a seminar on listening. Such seminars are frequently offered by local universities or technical colleges. In addition, some organizations specializing in management development also offer programs on improving listening effectiveness.

In addition to listening closely to what employees have to say, a technical manager also needs to deal with problems that are brought to his attention by staff members. When employees bring problems to the attention of their supervisors, they expect some type of response in a reasonable time frame. Ideally, an employee who brings up a problem would prefer to have her manager solve the difficulty. However, in some cases, solving the problem may not be practical or possible. In such cases, it is important for a technical manager to get back to the employee within a reasonable time frame and explain why a particular problem cannot or will not be solved.

For example, a research scientist might feel that the current computer hardware or software being used does not allow him to perform certain job tasks as rapidly as possible. As a result, the employee might ask his supervisor for new equipment or software. If the organizational unit is losing a great deal of money, purchasing this requested new hardware or software may not be possible. As a result, the manager of the unit needs to explain to the employee why the equipment or software cannot be purchased. (This manager might suggest the alternate approach of acquiring used equipment that, although not new, may still be better than that which is currently being used.)

A mistake that many technical managers make is that they fail to respond to an employee's concern in a reasonable time frame. In some cases, a manager may not respond because she is simply too preoccupied with other activities. In other cases, however, a manager may not respond because she is reluctant to tell the employee that the problem raised cannot be solved. In my experience, I have found that getting back to employees on a timely basis is essential. If an employee brings up a problem, he obviously would prefer to have that problem solved. However, I have found that most employees are willing to accept not having a problem solved as long as they receive a plausible explanation within a reasonable time frame. What virtually all employees find unacceptable is to hear nothing from a manager in response to a problem that has been brought to his attention.

10.9 COMMUNICATING THOROUGHLY, CANDIDLY, AND PROMPTLY

Technical employees have a need to know information that affects their ability to do their job as well as information about the department and the organization as a whole. If a manager of technical employees shared all of the information that her staff members would like to know, she might need to spend 100 percent of the time communicating, with no time for any other activities. Obviously, this would not be feasible. However, most technical managers probably could and should spend more time communicating with employees. Most employees prefer to get information about their unit or the company as a whole from their supervisor. As a result, communicating needs to be a high-priority activity for all technical managers [7].

Communication can be done in many ways. These include one-on-one discussions, staff meetings, electronic mail, and bulletin board postings. Regardless of the method used, however, it is essential that the communication be thorough and candid as well as timely. First, regarding promptness, a manager needs to ensure that he communicates information on a timely basis. For example, most employees would not be happy if their supervisor told them about the elimination of a number of the company's products after they already heard about this through the "rumor mill" and read about it in the newspaper. Often it is difficult for a manager to provide employees with timely information because of other priorities. Nonetheless, if significant information needs to be shared with employees, a technical manager simply needs to make time regardless of other pressing priorities. When employees do not receive timely information from their supervisors, they often feel neglected, and they may feel that their supervisors do not consider them to be integral parts of their organization or unit. When technical employees feel this way, they typically are not motivated to perform to the best of their abilities.

Another key aspect of effective communication is thoroughness. A technical manager should not waste her or staff members' time by providing unnecessary detail in communication. However, technical managers do need to be complete when they communicate. For example, suppose an organization makes home fetal monitors, 25 percent of which are sold overseas. In addition, suppose that a decision was made to halt all sales of monitors outside the United States. It would probably be ineffective for a technical manager to communicate this via a one-sentence memo on the bulletin board. In such a case, employees would probably have many questions about the reasons for the decision and the effect upon their jobs. As a result, a more appropriate communication vehicle for such a message might be a staff meeting. At such a meeting, a manager could make a thorough presentation concerning the decision and any questions that employees might have could be answered.

Another key element of communication is candor. A technical manager needs to be completely candid with employees at all times [1, 7]. Doing so builds one of the most important variables in manager/staff member relations: trust [7]. If an employee has complete trust in a technical manager, she is often motivated to go beyond the call of duty and do whatever it takes to get a job done. On the other hand, if a technical employee does not completely trust his supervisor, that individual's motivation to do the best job possible is not necessarily optimal. Technical employees are often fairly adept at detecting hypocrisy.

Some technical managers may not be completely candid with employees for various reasons. For example, some technical managers may be reluctant to share certain sensitive information with employees because they are afraid that the employees may not deal with the information in a responsible fashion. Other technical managers may be reluctant to share certain negative information because they are afraid that employees may respond in some adverse fashion. I feel that despite the potential drawbacks associated with complete candor, these are greatly outweighed by the significant advantages of such an approach.

Naturally, if top management has indicated that certain information is confidential and cannot be shared with employees, a technical manager needs to comply. In such a situation, if an employee asks about the confidential information, the manager could make a statement such as, "I'm sorry, but I really cannot discuss that." However, the manager should never say anything that is not true; otherwise, the employee's trust in the manager is placed at risk.

10.10 ALLOWING CONSIDERABLE AUTONOMY

Another key motivator that a technical manager can use to stimulate excellence in performance among her employees is to allow considerable autonomy [11]. Most technical employees appreciate being given a great deal of latitude in how they do their job. As a group in general, technical employees typically are bright individuals who are rather knowledgeable about certain areas. As a result, most of them appreciate and expect a fair amount of autonomy regarding how they do their jobs. In many cases, a technical employee is the inhouse expert in certain areas. As a result, it certainly makes sense to rely heavily upon such an individual's judgment with regard to his areas of technical expertise.

Some technical employees may not demonstrate a high need for autonomy. In such cases, a technical manager should work closely with these employees to build their confidence and encourage them to operate in a more autonomous fashion. By doing this, a technical manager helps to increase the effectiveness of employees, which in turn frees up some time for him that can be devoted to other high-priority activities.

To illustrate the allowance for and encouragement of autonomy by technical managers, we can use the following example. Claudine is a fairly new consulting engineer who works for John, a consulting engineering manager. Since Claudine just recently started working for the consulting firm and does not have previous experience, she might need, and perhaps even request, a great deal of direction.

John might use a very hands-on, directive approach with Claudine for the first few months of her employment. However, he probably should gradually allow and even encourage Claudine to exercise an increasing amount of autonomy as she gains experience.

For example, initially Bill might meet with some clients along with Claudine to assess the clients' needs. Then, Bill might provide Claudine with a fair amount of direction as to how the clients' needs could be met. Claudine might also spend a fair amount of time observing and talking to Bill while he works with some of his clients. In addition, Bill might give Claudine quite a bit of direction regarding her write-up of client projects.

However, as Claudine gains additional experience, it would be in her best interest and that of her supervisor for Bill to allow her to work significantly more autonomously. Thus, after six months or so, Claudine might meet with clients on her own, and she might meet with Bill only at certain key intervals (for example, at the beginning of the project, halfway through, and at the end of the project). By using this approach, Bill is likely to provide optimal motivation to Claudine to demonstrate outstanding performance. Such an approach is also likely to develop Claudine's confidence and to increase Bill's effectiveness in his job.

10.11 SUMMARY

This chapter discussed additional key motivators that a technical manager needs to use to attain and to maintain top-level performance in her work group. These motivators include: (1) providing interesting and intellectually stimulating work; (2) giving appropriate feedback; (3) rewarding top performance; (4) eliminating poor performance; (5) using people's talents; (6) building a true team; (7) listening and dealing with problems; (8) communicating thoroughly, candidly, and promptly; and (9) allowing considerable autonomy. The payoff from using these motivators is greatly enhanced individual and group performance.

References

[1] Quick, Thomas, *The Manager's Motivation Desk Book*, New York: Wiley, 1985.

[2] Skopec, Eric W., *Communicate for Success*, Reading, MA: Addison-Wesley, 1990.

[3] McCoy, Thomas T., *Compensation and Motivation: Maximizing Employee Performance with Performance-Based Incentive Plans*, New York: AMACOM, 1992.

[4] Mindell, Mark, and Gorden, William, *Employee Values in a Changing Society*, New York: AMACOM, 1981.

[5] Nash, Michael, *Making People Productive*, San Francisco: Jossey-Bass, 1985.

[6] Morgan, Clifford T., and King, Richard A., *Introduction to Psychology*, Fourth Edition, New York: McGraw-Hill, 1971.

[7] Nicholas, Ted, *Secrets of Entrepreneurial Leadership: Building Top Performance Through Trust and Teamwork*, Chicago: Enterprise, 1993.

[8] Shank, James H., *Team-Based Organizations: Developing a Successful Team Environment*, Burr Ridge, IL: Irwin, 1992.

[9] Osborn, A. F., *Applied Imagination*, Third Edition, New York: Scribner's, 1963.

[10] Manz, Charles, and Henry P. Sims, *Business Without Bosses: How Self-Managing Teams Are Building High-Performance Companies*, New York: Wiley, 1993.

[11] Fink, Stephen L., *High Commitment Workplaces*, New York: Quorum, 1992.

Chapter 11

Minimizing Turnover

11.1 FOLLOWING THE PREVIOUSLY MENTIONED SUGGESTIONS

If a technical manager wants to minimize turnover and retain his outstanding staff members, he must follow the various suggestions mentioned in previous chapters. For example, to minimize turnover, a technical manager must attract outstanding candidates who are likely to remain with the organization. To do this, a technical manager needs to follow the various suggestions mentioned in Chapter 1, which deals with the recruitment of employees. For example, to attract excellent people who are likely to stay with the organization, a technical manager needs to pay very well and he needs to ensure that the organization provides an excellent benefits package to employees. In addition, a technical manager needs to ensure that his organization has developed an excellent reputation for having outstanding products/services and employees. Using techniques such as realistic job previews and employee referral programs can also aid a technical manager in recruiting and retaining outstanding people. Using networking, professional organizations, and internship programs are additional ways that a technical manager might recruit outstanding individuals who are likely to stay with the organization.

To retain outstanding employees, a technical manager also needs to follow the various suggestions mentioned in Chapter 2, which deals with the selection of outstanding people. For example, when a technical manager sets the selection standards for a position, she should include the likelihood of staying with the organization as one of the standards used to evaluate candidates. In following the premise that the best predictor of future behavior is past experience, a technical manager should probably look for people who have had substantial tenure with previous organizations. An individual who appears to be an outstanding performer but who has been with four different organizations in the past six years is probably not likely to stay with the technical manager's organization for a considerable length of time. If a candidate for a technical position is someone who has just recently gotten out of school and has not had any nontemporary full-time positions, a technical manager needs to look for other factors that might suggest that a candidate is likely to stay with her organization. For example, an employee who has returned to the same employer for a number of consecutive summers during school might be likely to remain with a technical manager's organization.

I also made a suggestion in Chapter 2 that a technical manager needs to ensure that there is a good match between a candidate and the job specifications, the technical manager's preferences/expectations, and the departmental/organizational culture. Obviously, if there is a close match between a candidate and these three factors, it is much more likely that the candidate will remain with an organization than if there is not a good match.

Likewise, a technical manager needs to follow the various suggestions mentioned in Chapter 3, which deals with interviewing, to increase the probability of selecting someone who will remain with the her organization for a significant amount of time. Similarly, if a technical manager follows the various suggestions mentioned in Chapter 4, which covers the use of other selection devices such as tests, work samples, reference checks, and assessment centers, she can increase the likelihood of selecting an outstanding employee who will remain with the organization for a significant period of time. In addition, if a technical manager uses the information presented in Chapter 5, which deals with key selection criteria for technical employees, she can increase the probability of selecting a top-notch employee who will stay with the organization.

Chapter 6, which deals with one-on-one development, also contains quite a number of suggestions that, if followed, can help to increase the tenure of technical employees. All of the various techniques discussed in Chapter 6 are related to the improvement of technical employees' capabilities and performance. When technical employees are performing more effectively in their jobs, they tend to feel better about themselves, their job, and their organization. As a result, the probability of them remaining with their current employer is increased.

For example, if a technical manager spends a considerable amount of time coaching a particular technical employee, this is likely to have a beneficial effect upon the employee's performance. When such an employee is performing more effectively, it is likely that his self-esteem will increase and that the employee will probably also like his job even more. In addition, if a manager has taken a great deal of her time to coach a technical employee, the relationship between the two individuals is likely to be strengthened. The net effects of these various factors is that the technical employee is probably more likely to remain with his organization rather than seeking new career opportunities elsewhere.

Chapter 7, which deals with development in groups, also contains some suggestions that, if followed, can aid a manager in minimizing turnover among technical employees. The rationale that was described previously to explain how one-on-one development helps to minimize turnover also applies to development in groups. As technical employees improve their skills via inhouse training, they tend to feel more positively about themselves, their job, and the organization. As a result of this, they are probably less likely to leave their organization to pursue other career opportunities.

Other developmental techniques, which are described in Chapter 8, can also serve to aid a technical manager in reducing turnover among her employees. For example, if a technical manager spends time talking to employees about career development, this is an indication that she cares about the employees' futures in the organization. When technical employees believe that their future career is important to their supervisor, they are less likely to feel that they need to go elsewhere to pursue career opportunities.

When a manager of technical employees employs the two critical motivators described in Chapter 9, she can also help to minimize turnover among her technical employees. First, when a manager sets challenging but realistic expectations, a technical employee is likely to be successful in meeting his objectives. Obviously, a successful employee is much less likely to leave his organization than one who is not successful. In addition, when a manager demonstrates true concern for her staff members, this can also dramatically decrease the turnover among technical employees. When a manager shows that she truly cares about the employees' welfare, they feel that they are important to the manager and the organization and, hence, are much less likely to want to leave the organization in search of other career opportunities.

When a manager employs other key motivators, which are discussed in Chapter 10, turnover among technical employees can be minimized. All of the various motivators described in Chapter 10 serve to meet technical employees' needs. When employees' needs are being met, they are happy and have no reason to consider career opportunities in other organizations.

11.2 TREATING PEOPLE WITH RESPECT

As indicated earlier, following the various suggestions described in the previous chapters can help a manager to minimize turnover among his technical employees. In addition, one factor that can have a dramatic effect upon minimizing turnover among technical employees is treating staff members with respect.

People do not like being ignored or being treated condescendingly by their supervisor. Conversely, people enjoy being treated with respect [1–3]. When a manager of technical employees is able to convey his feeling of respect for his staff members, this is viewed very favorably by them. Dessler [2] reported that a leadership style that reflects respect for staff members is consistently correlated with high morale among employees.

Champagne and McAffee [1] stated that one indication of a supervisor's respect for his employees is that he treats them fairly. When they are treated fairly, employees tend to like their manager. Supervisors who are well liked are often able to increase their ability to exercise influence and authority with staff members [1]. One way in which they can influence staff members' behavior concerns turnover. In other words, individuals who feel that they are being treated fairly and with respect by their supervisor are much less likely to look for employment elsewhere than those who do not feel this way.

When technical employees feel that they are not being treated with respect, they typically do not feel any great allegiance to their supervisor or their organization. For example, one technical manager with whom I am familiar is a very arrogant, narcissistic individual. He treats others in the organization who are not at as high an organizational level as being "beneath" him. He tends either to ignore staff members or to deal with them in a very condescending fashion. I have heard him make comments such as the following to one of his staff members, "I don't have time to deal with insignificant issues like that. I have more important things to worry about." As a result of his actions and comments, this individual's staff members do not feel that he respects them. Because of this, they do not feel a very strong loyalty to him or to the organization, which has resulted in the exodus of many of his staff members to pursue other career opportunities.

Another technical manager whom I know treats his staff members with a great deal of respect. He considers his staff members to be valuable, knowledgeable professionals, and he treats them accordingly. One of the things he does to show respect is to treat all of his staff members fairly. He also has told them that if they have anything that they would like to discuss with him, regardless of whether it is something of major importance or not, he is very willing to talk to them about it.

Another way in which he demonstrates his respect for his staff members is that he deals with them in a manner that makes them feel that they are professionals and colleagues, not individuals who are beneath him. It should come as no surprise that none of this manager's staff members have resigned in the past ten years.

11.3 SUMMARY

The various ideas described in previous chapters can be very helpful to a manager in minimizing turnover among his technical employees. In addition, it is essential for a technical manager to treat his employees with respect. Although nothing that a manager does can *absolutely guarantee* that none of his employees will decide to leave the organization, by showing staff members true respect, a technical manager can have a substantial impact on minimizing employee turnover.

References

[1] Champagne, Paul J., and R. Bruce McAffee, *Motivating Strategies for Performance and Productivity: A Guide to Human Resource Development,* New York: Quorum, 1989.

[2] Dessler, Gary, *Improving Productivity at Work: Motivating Today's Employees,* Reston, VA: Reston, 1983.

[3] Quick, Thomas, *The Manager's Motivation Desk Book,* New York: Wiley, 1985.

Chapter 12

Putting It All Together

12.1 INTRODUCTION

Technical managers can use the information presented in this book in two ways. First, some technical managers may simply try to identify a few techniques described in the book that might be beneficial in their organization. Such an approach might well be helpful, but it is not ideal.

Other technical managers may use the information presented in a much more effective manner. Specifically, technical managers would do an assessment of how they currently handle people in the organization, involving all of the key areas described in this book: recruiting, selecting, developing, motivating, and retaining technical people.

After completing this evaluation, a technical manager would establish a plan for implementing *all* of the various techniques described in the book that would aid him in recruiting, selecting, developing, motivating, and retaining outstanding employees.

To illustrate the approach described above, consider a case study involving a fictional company, XYZ Software, Inc. This organization manufactures software for personal computers and has a staff of about 500, many of whom would be classified as technical employees. The vice president of technology is John Doe.

John feels that the company is performing adequately, but not spectacularly. Although he feels he has some good people, he admits that his evaluation of the technical employees as a group would place them in the category of "average." John feels there are a number of actions that he can take to enhance the quality of the technical employees and to improve the overall effectiveness of the organization. John recognizes that upgrading the quality of his employees from average to outstanding will be a complicated, time-consuming, and expensive endeavor.

However, changing a department and/or organization is not simple. It took a long time for XYZ Software, Inc., to develop an adequate workforce; changing this to an outstanding workforce cannot be done overnight.

12.2 RECRUITING OUTSTANDING EMPLOYEES

The first area that John Doe evaluated is the recruitment of technical employees. John's assessment concerning recruiting is that XYZ Software, Inc., is doing only a mediocre job in this area.

One key area relevant to the recruitment of outstanding people is compensation. To assist in evaluating the company in this area, John recommended using a compensation consultant. After reviewing the results of the compensation survey done by the consultant, John decided that the money paid to technical employees is slightly below average. John recognizes that paying slightly below the market rate is unlikely to attract many outstanding employees. As a result, he decided to change the compensation policy for technical employees so that the midpoint of each position's salary range is at the 90th percentile compared with the market. Naturally, if the average technical employee in the company is in the top 10 percent of pay in comparison to the industry as a whole, the average XYZ Software, Inc., employee must also perform in the top 10 percent as compared to the entire industry. This requirement is necessary for existing as well as new employees.

Another important consideration regarding recruiting that was mentioned in Chapter 1 is the benefits package. After evaluating XYZ Software, Inc.'s benefits package, John Doe decided that several of his company's benefits are rather weak compared to those offered by other organizations. As a result, John decided to recommend to the president of XYZ Software, Inc., that the company's benefits package be improved so that it compares favorably with other organizations that are competing for the same employees. The increase in cost associated with the improved benefits package will be covered in the same manner as the increase in compensation cost will be handled: by the concomitant increase in performance/productivity resulting from the recruiting of a significantly better caliber of employee.

John Does also evaluated the company's reputation for products and employees. After getting some candid feedback from a number of different sources, John determined that the organization's reputation is merely adequate. As a result, John decided that he needs to take various actions to improve the organization's reputation for its products and employees. Improving the compensation and benefits is expected to have a dramatic effect upon the overall quality of employees, which, in turn, should have a significant effect upon the company's reputation concerning its employees. To improve the organization's reputation with respect to its products, John recommended to the president that the organization adopt a new strategy. The company has been developing only one or two brand new products per year in addition to current product updates. The new strategy involves focusing on annually developing three to four leading-edge new products as well as updating existing products. If XYZ Software, Inc., is successful in implementing this new strategy, its reputation concerning its products is likely to improve. In addition, its bottom-line results are also likely to become more favorable.

An additional aspect of the recruiting process that John evaluated concerns the various techniques that the organization is using for recruiting. John's assessment of the company's recruiting techniques indicated that only mediocre marks are deserved in this area. The company has been recruiting its employees primarily through local newspaper advertisements. Very few outstanding people have been found via this approach.

To improve the company's performance in recruiting, John decided to implement a formalized employee referral program concerning prospective employees. He also decided to begin to use realistic job reviews with candidates and plans to implement an internship program. John also determined that he will make significantly better use of the various contacts in professional organizations. Finally, John decided that he will identify and use effective employment agencies and executive recruiters when appropriate.

12.3 SELECTING EXCELLENT PEOPLE

The selection of technical employees is another area that John has reviewed. First, with regard to selection standards, John determined that these generally are either nonexistent or not particularly well-specified. As a result, John decided to ensure that the selection standards for each position are very specific. In addition, he determined that the selection standards in general need to be raised to a high but realistic level.

John has also assessed the match between the candidates for various open positions and the job specifications, the supervisory preferences/expectations, and the departmental/organizational culture. His evaluation in this area indicated that the match in these areas has not always been particularly good, and in many cases an assessment of the match has not even been done. As a result, John decided that he will encourage all the technical managers to use the questions indicated in Chapter 2 to determine the quality of the match between the candidates and the jobs, the supervisory preferences/expectations, and the departmental/organizational culture.

Another area that John evaluated with regard to the selection of technical employees is that of interview planning. His assessment of this area indicated that it has been mediocre at best, and nonexistent in many cases in the past. As a result, John decided he will request that all of his technical managers spend an adequate amount of time in interview planning. This involves, for example, that all technical managers use standard as well as tailored questions, as described in Chapter 3.

Many of the questions provided for the product design engineer position described in Chapter 5 can also be used for some of the company's technical positions.

Selection interviews is an additional area that John reviewed. His evaluation determined that most selection interviews in the past have been brief and unstructured. As a result, John will expect all of his technical managers to do structured, in-depth selection interviews in the future.

John also assessed the interpretation of interview information. John's evaluation indicated that many technical managers are not comfortable and do not feel competent to interpret data from interviews. As a result, John decided to bring in an outside consultant who is an expert in interviewing. This individual will provide training to technical managers on effective interviewing in general, including the interpretation of interview data.

Another area that John reviewed is that of selection techniques other that interviews. After evaluating this, John determined that considerable improvement is needed. He determined that reference checking is generally only mediocre. As a result, John plans to urge all of his technical managers to use the reference checking format suggested in Chapter 4. John also determined that no tests are currently being used in the selection process. As a result, he plans to use the services of a qualified consultant, who will do in-depth job analyses in key positions. The consultant will also identify/develop job-relevant tests to help improve the selection of people for these key jobs.

12.4 DEVELOPING OUTSTANDING EMPLOYEES

12.4.1 One-on-One Development

An additional major area that John Doe reviewed is the development of technical employees. After looking at the one-on-one development of the technical employees who work for XYZ Software, Inc., John decided that a great deal of improvement is needed. First, he decided that all technical managers need to help their employees define their developmental needs and to assist them in identifying vehicles to meet these needs.

John also plans to encourage all of his managers to coach and to counsel their technical employees. Additionally, he also plans to urge them to set up ability development teams to assist in the developmental process. Finally, John plans to use 360-degree feedback programs (that is, ones involving feedback from coworkers) such as the MADI and the EADI programs described in Chapter 6.

12.4.2 Inhouse Training

After evaluating the inhouse training that is conducted at XYZ Software, Inc., John Doe determined that this area is mediocre at best. To remedy the situation, John plans to assure that an adult learning model is used for all inhouse training sessions. He will also ensure that multiple-source training needs analyses are used to provide all of the necessary information concerning the topics that need to be addressed in the training programs.

Additionally, John plans to encourage all the managers in his area to meet with their employees prior to and following their participation in classes. He also plans to ensure that all his managers expect their employees to complete developmental action plans after they have completed courses.

Another change that John plans to implement concerns the development and implementation of an internal program designed to teach new employees about XYZ Software, Inc.'s culture. Finally, John plans to ensure that all courses taken by his employees are evaluated using the four types of measurement described in Chapter 7 (that is, reactions, learning, behavior change, and organizational performance improvement).

12.4.3 Other Development

After assessing the other development that has been going on in his department, John determined that this, too, is mediocre at best. To remedy the situation, John plans to encourage his employees to take outside courses. He also plans to encourage the technical managers in his area to establish formal reading programs for their employees to foster their development.

In addition, John decided to implement formal career development and succession planning programs in his department. Finally, he determined that he will use the services of a qualified outside consultant to assist in the area of team building for his department.

12.5 MOTIVATING OUTSTANDING EMPLOYEES

John Doe evaluated the level of motivation within the technology department, and he determined that it is merely average. To change this situation, he decided to ensure that he sets challenging but realistic expectations for the managers who report directly to him. He also plans to encourage them to do the same with their employees. In addition, John decided to ensure that the employees in his department understand that he truly cares about their well-being. He might demonstrate his concern for his employees, for example, by taking the time to ask them about their families and hobbies. Another example of how John might demonstrate his true concern for his employees would be to send get well cards and visit employees or their family members who are seriously ill.

John also plans to improve the motivational climate in his department by ensuring that employees are given regular, appropriate feedback concerning their performance. This includes frequent informal, positive feedback and informal constructive criticism when needed. It also includes regular, in-depth formal performance reviews of all staff members.

Another action that John plans to take to increase the technical employees' motivation is to be certain that top performance is rewarded appropriately. This might involve, for example, providing exceptional cash bonuses to those employees who achieve outstanding results.

Additionally, John Doe plans to improve the overall motivation of technical employees in his unit by making every effort to eliminate poor performance. Regarding this, he might, for example, ensure that any technical employees who are not performing up to standards do not receive merit increases. In addition, he might indicate that any technical employees who do not perform up to standards after being given sufficient assistance will not continue in their present position within the organization.

Another key motivator mentioned in Chapter 10 that John plans to ensure is used in his organization is using employees' talents. Particularly, he might make certain that all of the technical employees are given as much autonomy as possible in doing their jobs. John might also encourage his managers to have formal brainstorming sessions. Such meetings might allow all employees to provide their input concerning, for example, new product ideas and ways to deal with existing problems.

John also plans optimize the motivational climate in his department by making every effort to promote a strong feeling of camaraderie among his staff members. Conducting formal team-building sessions using an outside consultant, as was mentioned previously, would help in this regard. In addition, John might be certain that he mentions that he expects his staff members to cooperate rather than compete with one another. Highlighting cases in which employees have done an effective job of cooperating with other team members might be very helpful in this regard.

In addition, John plans to ensure that the managers in his department are listening and dealing with employees who perceive problems. One way that John might be able to do this is to conduct periodic employee surveys. These provide staff members with an opportunity to highlight any problems that they perceive. After evaluating the results of such surveys, it is essential that John follow up to ensure that all of the problems that are mentioned in each survey are addressed. This does not necessarily mean that he needs to take specific action with regard to every single problem. It does mean, however, that in cases where certain problems cannot or will not be solved, the employees will be given an explanation as to why these problems cannot or will not be addressed.

John also plans to be certain that the motivational climate is optimal by verifying that all communication is thorough and candid. Obviously, John can ensure, first of all, that all of his communication with employees is thorough and candid. John might also occasionally sit in on staff meetings conducted by those who report directly to him to make certain that their communication is also effective. In addition, he might review some of the written communication done by his managers periodically to ensure thoroughness and candor.

12.6 RETAINING EXCELLENT EMPLOYEES

Another key area that John has evaluated is that of employee retention. His assessment indicated that the turnover of his department is unacceptably high. Taking the various actions described previously is likely to have a favorable impact upon employee retention. In addition, John plans to make sure that all employees in his department are treated with respect. In this regard, he plans to evaluate all of the policies and procedures used in his department to ensure that none of these causes his employees to feel that they are not respected. For example, John might review the customary practices concerning employee dismissal. With regard to this area, he needs to be certain that when employees are being dismissed from their jobs, they are still able to retain their dignity and self-respect—as a result of being treated as humanely as possible.

12.7 SUGGESTED ACTIONS FOR TECHNICAL MANAGERS

I strongly recommend that every technical manager evaluate the recruitment, selection, development, motivation, and retention of employees in her organization, just as John Doe did in the fictional case study. I suggest that technical managers outline the key points made in each of the chapters. This outline can then be used to evaluate their departments in various areas. In addition, the outline may be helpful in highlighting possible corrective actions that can be taken to deal with any problem areas that are identified.

12.8 SUMMARY

This chapter presented a fictional case study to illustrate how a technical manager might evaluate the recruitment, selection, development, motivation, and retention in his organization. The fictional case study also illustrates some of the corrective actions that a technical manager might take to deal with any areas of deficiency that are noted.

Recruiting, selecting, developing, motivating, and retaining excellent employees can be very challenging. I believe that I provided a number of practical suggestions in this book that can assist a technical manager considerably in performing these tasks. Following the various suggestions in this book will not guarantee success, but doing so will increase a technical manager's effectiveness in his job.

Appendix A

Additional Information Relevant to Chapter 2

A.1 KSAPs

The relationship between knowledge, skills, abilities, personal attributes, and job specifications can be demonstrated using the following equation: $K + S + A + P = JS$.

Knowledge (K) can be defined as information on a particular job-relevant topic. For example, a manufacturing engineer might need to have a general knowledge of various manufacturing processes.

A skill (S) is the capability to perform a job-relevant task using some type of equipment. For example, a product designer might need to have the skill to use computer-assisted design hardware and software to design products.

An ability (A) can be defined as the capability to perform some job-relevant task that does not involve using equipment. For example, a consulting engineer might need the ability to make effective oral presentations to clients and prospective clients.

A personal attribute (P) is a personality characteristic that is relevant to performing a job-relevant task. For example, a chemist might need to posses the personal attribute of meticulousness to attain the necessary precision in mixing various chemicals while conducting experiments.

Knowledge in several different areas might be necessary for a given job. Likewise, multiple skills, abilities, and personal attributes might be relevant to a particular job. When all of these are added together, they comprise the job specifications (JS). The job specifications indicate what is needed to do a job effectively.

A.2 Knowledge, Skills, Abilities, and Personal Attributes (KSAPs) Needed to Perform Effectively in a Particular Position

1. To perform effectively in this position, what knowledge is:
   a. Absolutely critical?
   b. Extremely important?
   c. Very helpful?
2. To do this job well, what skills are:
   a. Absolutely critical?
   b. Extremely important?
   c. Very helpful?
3. To perform effectively in this role, what abilities are:
   a. Absolutely critical?
   b. Extremely important?
   c. Very helpful?
4. To do this job well, what personal attributes are:
   a. Absolutely critical?
   b. Extremely important?
   c. Very helpful?

A.3 Supervisory Preferences/Expectations That Must Be Taken Into Account Regarding Candidates for a Particular Position

To perform effectively in this position, what supervisory preferences/ expectations are:

a. Absolutely critical?
b. Extremely important?
c. Very helpful?

A.4 Aspects of the Departmental and/or Corporate Culture With Which a Candidate Must Be Able to Cope Effectively to Perform Competently in a Particular Position

1. It is absolutely critical that a candidate for this position be able to cope effectively with which aspects of the departmental and/or corporate culture?

2. It is extremely important that an individual considered for this job be able to deal effectively with which aspects of the departmental and/or corporate culture?

3. It is very helpful that a candidate for this role be able to cope effectively with which aspects of the departmental and/or corporate culture?

A.5 KSAPs Needed to Perform Effectively in the Position of Manufacturing Engineer for the Acme Pen Company

1. Knowledge
   a. Absolutely critical knowledge: state-of-the-art plastics and metals manufacturing processes;
   b. Extremely important knowledge: current writing instrument manufacturing processes;
   c. Very helpful knowledge: manufacturing processes used currently at Acme.

2. Skills
   a. Absolutely critical skill: using a computer to design production equipment;
   b. Extremely important skill: operating production equipment that is currently used at Acme;
   c. Very helpful skill: word processing.

3. Abilities
   a. Absolutely critical ability: problem solving;
   b. Extremely important ability: mechanical;
   c. Very helpful ability: creativity.

4. Attributes
   a. Absolutely critical personal attribute: conscientiousness;
   b. Extremely important personal attribute: capacity for relating to others;
   c. Very helpful personal attribute: perseverance.

A.6 John Doe's Preferences for/Expectations of Candidates for the Manufacturing Engineer Job

1. Absolutely critical preference/expectation: willingness to take direction;

2. Extremely important preference/expectation: willingness to communicate regularly;

3. Very helpful preference/expectation: capacity to adhere to budget.

A.7 Aspects of the Culture of the Manufacturing Engineering Department and/or Acme Pen Company With Which a Candidate for the Manufacturing Engineer Position Must Be Able to Cope Effectively

1. Absolutely critical: strong sense of urgency;
2. Extremely important: good work ethic;
3. Very helpful: team cooperation.

Appendix B

Additional Information Relevant to Chapter 3

B.1 SOME SAMPLE INTERVIEW QUESTIONS

B.1.1 Work Experience

1. a. What interests you most about the job for which you are interviewing?
   b. What aspects of the job you are being considered for are not ideal?

2. (These questions all pertain to the candidate's present or most recent job.)
   a. What do you enjoy most?
   b. What are your greatest accomplishments?
   c. What have you enjoyed least?
   d. At what times do you usually start and end work each day?
   e. How many hours do you typically work weekly on weekends and in the evening?
   f. What was your starting compensation and what are you currently making (or what was your most recent salary)?
   g. For what have you received praise?
   h. What areas needing improvement have been mentioned?
   i. How has your performance been rated?
   j. Why are you considering leaving (or what were some factors that influenced your decision to leave)?

3. (For each of the candidate's previous three positions, ask the same questions indicated in Question 2.)

4. (For all positions held before the previous three jobs, ask the candidate about the main reasons for leaving each one.)

B.1.2 Education

1. a. What was your G.P.A. in high school?
   b. How much did you study in high school compared to others?
2. (Ask the two questions above regarding college/technical college.)
3. (Ask the first two questions above regarding graduate school, if relevant.)
4. How many hours did you work per week during the school year while you were in each of the following:
   a. High school?
   b. College/technical college?
   c. Graduate school (if relevant)?
5. In what outside activities did you participate while you were in each of the following:
   a. High school?
   b. College/technical college?
   c. Graduate school (if relevant)?
6. How bright are you compared to others?

B.1.3 Other

1. What types of people have you enjoyed working with most?
2. What types of individuals have you enjoyed working with least?
3. What do you expect from your supervisor?
4. What do you expect from your employer?
5. What is your management style (if relevant)?
6. What is the relative emphasis on results versus people in your management style (if relevant)?
7. a. How would you rate yourself as a manager (on a 1 to 10 scale) (if relevant)?
   b. What evidence can you provide to support your rating?
8. What is your decision-making style?
9. a. How would you rate yourself on assertiveness (on a 1 to 10 scale)?
   b. What evidence supports your rating?

10. How would you rate yourself regarding the technical areas of the job (on a 1 to 10 scale) and what evidence can you provide for each of the ratings?
11. What changes have you noted in yourself in the last few years?
12. a. To summarize, what are your greatest assets or strengths regarding the position for which you are interviewing?
    b. To summarize, in what areas do you need some development regarding the job for which you are being considered in this organization?

B.2 APPROPRIATE AND INAPPROPRIATE INTERVIEW QUESTIONS

Directions: Indicate whether each question would be appropriate or inappropriate in an interview.

1. How did you learn to speak Spanish so well?
2. Marquette University is a Jesuit school, isn't it?
3. How old are your kids?
4. Do you have any physical disabilities?
5. Have you ever been arrested?
6. Would you be willing to relocate?
7. Do you have any reason that would prevent you from traveling when required?

B.3 INFORMATION TO HELP EVALUATE RESPONSES IN SECTION B.2

1. Inappropriate. (It might indicate that a person is of Mexican origin and could be used to discriminate against a candidate.)
2. Inappropriate. (Questions concerning the religious affiliation of a school might cause an interviewer to draw a conclusion about a candidate's religion and could be used to discriminate.)
3. Inappropriate. (This question could be used to determine whether an applicant's care of young children might interfere with job responsibilities and could be used to discriminate.)
4. Inappropriate. (This could be used to discriminate and is specifically forbidden by the Americans with Disabilities Act.)

5. Inappropriate. (Arrests do not necessarily imply guilt, and this question could be used to discriminate against certain candidates.)
6. Appropriate if job-related.
7. Appropriate if job-related.

Appendix C

Additional Information Relevant to Chapter 4

C.1 A COMPARISON OF SELECTION DEVICES OTHER THAN INTERVIEWING

Selection Device	*Candidate's Time Involvement*	*Manager's Time Involvement*	*Consultant Needed*	*Cost*
Test	Varies (#1)	Not extensive	Yes	Varies (#2)
Work Sample	Varies (#3)	Not extensive	No	Minimal
Reference Check	None	15–30 minutes	No	Minimal
Assessment Check	Often 1–3 days	Varies (#4)	Yes	Significant

(#1) Often 30–60 minutes.
(#2) Often no more than a few dollars per test. Could be more if computerized scoring/interpretation is used.
(#3) Often between 15 minutes and several hours.
(#4) Could be extensive if a manager is an observer or administrator.

C.2 EXAMPLES OF WORK SAMPLES FOR CERTAIN SELECTED POSITIONS

Position	*Work Sample*
Engineering technician	Analysis/recommendations regarding building a prototype
Product design engineer	Analysis/recommendations regarding a product design problem
Manufacturing engineer	Analysis/recommendations regarding a problematic manufacturing process
Consulting engineer	Anaysis/recommendations regarding a client problem

C.3 SOME RECOMMENDED QUESTIONS/COMMENTS FOR REFERENCE CHECKING

1. a. Did _______________ report directly to you?
   b. How long did _______________ work for you?
2. On a scale of 1 to 10, how would you rate _______________'s overall job performance when the individual worked for you?
3. What were _______________'s major strengths?
4. a. What were _______________'s major developmental needs when the individual worked for you?
   b. What about (indicate other developmental needs identified by, for example, the interview and test results)?
5. Would you rehire _______________? (If not, ask for an explanation.)
6. (Describe the position for which the candidate is interviewing.) How do you think _______________ would do in this position as I have described it to you?
7. Thank you very much for your time. You've been extremely helpful.

Appendix D

Some Interview Questions for the Design Engineer Position in the Case Study in Chapter 5

1. a. How would you rate yourself regarding each of the technical aspects of the product design engineer position (on a 1 to 10 scale)?
   b. What evidence can you provide to support each of the above ratings?
2. a. What were some key accomplishments in your present (last) position?
   b. What were some key accomplishments in each of your previous positions?
3. What evidence can you provide that illustrates your ability to design pressure sensors for the auto industry?
4. a. Can you provide an example that illustrates your ability to work closely with manufacturing in the product design process?
   b. How do you feel about working closely with the manufacturing function?
5. Can you provide an example that illustrates your ability to manage various projects simultaneously?
6. a. How would you describe your management style?
   b. How would you rate yourself as a manager (on a 1 to 10 scale)?
   c. What evidence can you provide to support the above rating?
7. a. What time do you typically start and end your work day?
   b. How many hours do you typically work in the evening and on weekends?
   c. How many hours per week did you typically work during the school year while you were in college/technical college and graduate school (if relevant)?

8. Can you provide an example that illustrates your persistence?

9. What evidence can you provide that illustrates your conscientiousness?

10. a. How would you rate your self-confidence on a scale of 1 to 10?
    b. What evidence can you provide to support the above rating?

11. a. How would you rate yourself on assertiveness (on a 1 to 10 scale)?
    b. What evidence can you provide to support the above rating?

12. a. What was your G.P.A. in college/technical college?
    b. What was your G.P.A. in graduate school (if relevant)?
    c. How bright are you, as compared to others?

13. What evidence can you provide to illustrate your creativity?

14. a. How would you rate your oral communication ability (on a 1 to 10 scale)?
    b. What evidence can you provide to support the above rating?
    c. How would you rate your written communication ability (on a 1 to 10 scale)?
    d. What evidence can you provide to support the above rating?

15. a. What do you expect from your supervisor?
    b. What evidence can you provide that illustrates your ability to function autonomously?
    c. What evidence can you provide to illustrates your ability to take direction?

16. a. In summary, what do you view as the major strengths/assets that you could bring to this position?
    b. In summary, what do you view as some of the areas in which you could perhaps grow/develop with regard to this position?

Appendix E

Additional Information Relevant to Chapter 6

E.1 EFFECTIVENESS FEEDBACK FORM

To be completed on:
Return by __ /__ /__. Send to:

Instructions: I would appreciate receiving candid, anonymous feedback from you and some of my other coworkers concerning my work effectiveness. This information will be very valuable to me in identifying my strengths as well as areas that need some improvement. Please type your comments to help preserve your anonymity. Thank you for your help.

1. Concerning my overall work performance, what do you feel that I do well?

2. What aspects of my work performance need improvement from your perspective?

E.2 SOME SAMPLE MANAGEMENT ASSESSMENT AND DEVELOPMENT INVENTORY ITEMS

1. To what extent does this individual give positive feedback regarding effective performance? (motivating)
2. To what extent does is this individual adequately assertive? (relating)
3. To what extent is this individual effective in coaching and counseling others to help improve their performance? (developing)
4. To what extent does this individual encourage teamwork? (leading)

(Each item on the actual inventory has a five-point scale. The lowest rating is one, meaning to a very little extent, not at all, very infrequently or never (that is, estimated to occur 0 percent to 20 percent of the time, if frequency is relevant). The highest rating is five, meaning to a very great extent, totally, quite often, or always (that is, estimated to occur 80 percent to 100 percent of the time, if frequency is relevant). If the person who is doing the rating is unable to rate a particular item, it is marked "NA" (not applicable).)

(For more information on the Management Assessment and Development Inventory contact: Douglas M. Soat, Ph.D., Consulting Psychologist, 2600 N. River Bluff Drive, Janesville, WI 53545-0701; 608-756-1700).

E.3 SOME ITEMS FROM A SAMPLE MADI FEEDBACK REPORT

Coworkers' Ratings *1 2 3 4 5*	*Average Coworker Rating*	*Self Rating*	*Difference*
0 1 3 5 2	3.73	4.00	–0.27
Delegating	3.72	3.29	+0.43
Overall	3.70	3.49	+0.21

Comments

1. Strengths

- Manages people well; very organized; hard-working.
- Cooperative, creative, and responsible.

2. Improvements Needed

- Address problems openly and promptly.
- Distribute workload evenly; recognize achievements.

E.4 SOME ITEMS FROM A SAMPLE MADI DEVELOPMENTAL ACTION PLAN

1. Demonstrate a willingness to meet commitments to others.

- Return all phone calls within 24 hours, starting immediately.
- Respond to all communications within one week, starting immediately.
- By May 1, develop and implement a followup/reminder system to monitor all projects/deadlines.

2. Improve the dissemination of relevant information relating to my department to those within and outside of my area.

- Have weekly meetings with my staff, starting next Monday at 9 AM.
- Meet weekly with individuals who have daily contact with my department to discuss relevant issues, starting next week.
- Distribute copies of memos describing departmental actions to all individuals who desire this information, starting immediately.

E.5 SOME SAMPLE EMPLOYEE ASSESSMENT AND DEVELOPMENT INVENTORY ITEMS

1. To what extent does this individual avoid taking too long to make decisions? (decision making)
2. To what extent does this individual cooperate with others? (cooperating)
3. To what extent is this individual effective in getting things done? (doing)
4. To what extent does this individual tell others how he/she really feels? (relating)

(The rating scale is the same as that used with the MADI; see Section E.2. For more information on the EADI, contact the author at the address indicated in Section E.2.)

About the Author

Doug Soat, Ph.D., is a self-employed consulting psychologist in Janesville, Wisconsin. He provides psychological and human resources–related consulting services, including assessment/development of employees and organizations, to businesses and other entities. His clients include Burdick, Inc.; Electrol Specialties Company; Driv-Lok, Inc.; and numerous other organizations.

Dr. Soat has worked with technical employees for over 20 years. He has had substantial inhouse experience as a professional and/or a human resources executive at the vice presidential level in corporations including the Parker Pen Company, Sentry Insurance, and SSI Technologies, Inc., providing an excellent complement to his experience as an external consulting psychologist.

He received his B.A. degree in psychology from Cornell University in Ithaca, New York. His M.S. and Ph.D. degrees in psychology were earned at Marquette University in Milwaukee, Wisconsin. He received his M.B.A. degree in management at the University of Wisconsin in Whitewater, Wisconsin. He has written numerous articles for national business publications and has made various presentations to national technical organizations.

Index

The Artech House Technology Management and Professional Development Library

Bruce Elbert, *Series Editor*

Applying Total Quality Management to Systems Engineering, Joe Kasser

Engineer's and Manager's Guide to Winning Proposals, Donald V. Helgeson

Evaluation of R&D Processes: Effectiveness Through Measurements, Lynn W. Ellis

Global High-Tech Marketing: An Introduction for Technical Managers and Engineers, Jules E. Kadish

Introduction to Innovation and Technology Transfer, Ian Cooke, Paul Mayes

Managing Engineers and Technical Employees: How to Attract, Motivate, and Retain Excellent People, Douglas M. Soat

Preparing and Delivering Effective Technical Presentations, David L. Adamy

Successful Marketing Startegy for High-Tech Firms, Eric Viardot

Survival in the Software Jungle, Mark Norris

The New High-Tech Manager: Six Rules for Success in Changing Times, Kenneth Durham and Bruce Kennedy

For further information on these and other Artech House titles, contact:

Artech House
685 Canton Street
Norwood, MA 02062
617-769-9750
Fax: 617-769-6334
Telex: 951-659
email: artech@artech-house.com

Artech House
Portland House, Stag Place
London SW1E 5XA England
+44 (0) 171-973-8077
Fax: +44 (0) 171-630-0166
Telex: 951-659
email: artech-uk@artech-house.com

WWW: http://www.artech-house.com/artech.html